DATE DUE

The Challenge of Arctic Shipping

The Challenge of Arctic Shipping: Science, Environmental Assessment, and Human Values presents a collection of candid essays on the future of arctic waters. The editors and contributors address issues critical to arctic development, examining the implications of policy making in the north and its consequences for native people. The intricacies of decision making in an atmosphere of uncertainty are thoroughly explored, as is the significance of access to information, influence, and power.

Also examined are the activities and events associated with commercial proposals for the transportation of hydrocarbons through environmentally sensitive waters. The editors observe that the resulting political manoeuvring indicates that new approaches to this and other problems associated with the north are needed.

CYNTHIA LAMSON is associate director, Oceans Institute of Canada / Institut canadien des océans, Halifax, and assistant professor (research) at the School for Resource and Environmental Studies, Dalhousie University, Halifax.

DAVID L. VANDERZWAAG is director of the Marine and Environmental Law Program, Faculty of Law, Dalhousie University. He is also a research associate at the Oceans Institute of Canada / Institut canadien des oceans.

McGill-Queen's Series in Native and Northern Studies

Bruce G. Trigger, EDITOR

The Challenge of
Arctic Shipping:

Science, Environmental Assessment, and Human Values

Edited by
David L. VanderZwaag and
Cynthia Lamson

McGill-Queen's University Press
Montreal & Kingston • London • Buffalo

To the memory of J. Fielding Sherwood

© McGill-Queen's University Press 1990
ISBN 0-7735-0700-0

Legal deposit first quarter 1990
Bibliothèque nationale du Québec

Printed in Canada on acid-free paper

This book has been published
with the help of a grant from the
Social Science Federation of Canada,
using funds provided by the
Social Sciences and Humanities
Research Council of Canada.

Canadian Cataloguing in Publication Data
Main entry under title:

The Challenge of arctic shipping

(McGill-Queen's series in northern and native
studies; 2)
Includes index.
Bibliography: p.
ISBN 0-7735-0700-0

1. Shipping–Environmental aspects–
Arctic Ocean.
2. Environmental impact analysis–
Arctic Ocean.
I. VanderZwaag, David L.
II. Lamson, Cynthia   III. Series.

HE935.C43 1990   333.7'1'091632   C89-090236-4

Contents

Preface

Science and environmental assessment are much like religion. There are numerous variations in doctrine and practice. Persons may be converted or unconverted as to the value of such exercises. Depending on one's interests and priorities, the vision of the "right way" may differ and be subject to great debate. This book illustrates the diversity of perspectives on science and environmental assessment in the context of arctic shipping proposals and Canadian decision-making experience.

Part I examines the acquisition and application of arctic marine scientific knowledge. The link between shipping and scientific enterprise is described in chapter 1. Because the Arctic is a unique operational environment, three areas of scientific inquiry are particularly noteworthy: physical oceanography (chapter 2) marine mammal behaviour (chapter 3), and seabirds (chapter 4). Part II presents the conflicting visions of different participants in the assessment process: industry (chapter 5), Inuit (chapter 6), and territorial and federal governments (chapters 7 and 8, respectively). The three concluding chapters, in part III, suggest that decision-making on a project-by-project basis fragments science and inhibits understanding of ecosystem and social system dynamics. The authors argue that decisions about arctic shipping proposals must take into account general needs such as better risk assessment of hydrocarbon transport projects (chapter 9), consideration of the full spectrum of ocean management options for northern waters (chapter 10), and recognition that environmental assessment is an evolving process (chapter 11).

Support for this undertaking was generously provided by the Donner Canadian Foundation, which awarded funding to the Canadian Northern Waters Project of the Dalhousie Ocean Studies Programme (DOSP), at Dalhousie University in Halifax, to examine Arctic Ocean development and management issues over a four-year period. After the inception of the Northern Waters Project, DOSP amalgamated with the Canadian Marine Transportation Centre (CMTC) to become the International Institute for Transportation and Ocean Policy Studies (IITOPS), based in Halifax. In 1989 IITOPS was renamed the Oceans Institute of Canada / Institut canadien des océans. Previous project studies addressed questions concerning technology, resources, and regulatory capacity; this volume focuses explicitly on decision-making under conditions of uncertainty. Whereas the pace and scale of Arctic Ocean development have been checked since this book was conceived by its editors, we believe that *The Challenge of Arctic Shipping* raises important, basic questions relevant to resource development in general, for example, the use (and potential abuse) of science in environmental assessment and the capacity of decision-makers to comprehend and weigh competing, often irreconcilable values.

Special thanks to Judy Reade, research associate, School for Resource and Environmental Studies, Dalhousie University, for assisting in the final editing of this book. We also wish to thank Ena Morris, computer operator, Oceans Institute of Canada, Halifax, for word-processing services far beyond the call of duty.

Financial assistance from the Donner Canadian Foundation enabled the editors to prepare this book, and we are grateful for its generous support.

David VanderZwaag and Cynthia Lamson

Introduction
Gordon Beanlands

The explosion of interest in the environmental implications of arctic shipping can be traced to the convergence of a number of seemingly unrelated developments. First, the energy crisis of the early 1970s led to development of Canadian policies that promoted national self-sufficiency in the production of oil and gas. These, in turn, increased interest in frontier exploration, primarily in offshore and arctic regions, which resulted in the discovery of commercially viable reserves in the early 1980s. In most cases, shipment of crude oil or liquefied natural gas by ocean-going tanker was considered a major transportation option in the development of these reserves.

Second, at about the same time, environmental impact assessment (EIA) was coming of age. The requirement to take account of environmental and social consequences in the planning of economic development activities is an outgrowth of the global environmental movement, which led to the Stockholm Conference on the Human Environment in 1972. During the following few years many countries, including Canada, adopted the concept of EIA. At the federal level in Canada, EIA policy is articulated in the Environmental Assessment and Review Process (EARP), which was approved by cabinet in 1973. EARP applies to all agencies of the federal government and comes into effect if development projects involve federal land or money or result from federal initiatives. Thus, by the early 1980s, proponents of industrial developments in the Arctic encountered an environmental assessment process which had become a fairly sophisticated planning review mechanism.

Third, the growing demand of the aboriginal peoples of Canada for self-determination, in particular for control over their ancestral lands, added another complex dimension to the relationship between environment and development in the Arctic. Although the issue of aboriginal land claims had existed for some time, the potential environmental, social, and cultural effects of large-scale developments in the north, as well as potential economic gains, pushed the matter into the political arena. The implications were particularly important in Canada's arctic region, where the aboriginal peoples had never relinquished claim to the land (including offshore areas) through treaties. Thus development projects became checkers in a game of politics between aboriginal peoples and the federal government and at times have threatened to upset the environmental assessment process.

Standing as a backdrop to these relatively recent developments is the long history of marine navigation in the Arctic Ocean. Given the extreme difficulty posed for ships operating in such high latitudes, it is hard to comprehend the persistence in travelling by sea as opposed to land-based exploration and development. There are two obvious reasons for this, however. First, it was the search for a shorter shipping route to the Orient that motivated early arctic explorers. Most western European countries had built their empires on well-developed capabilities in marine transportation, and it was only natural to push this capability to its limits. Second, it was the riches of the arctic waters, particularly whales, that attracted the earliest commercial ventures and was the main reason for continued interest in the area up until the early years of this century.

This early focus on exploration and exploitation of biological resources gradually gave way to activities associated with military defence, delivery of government services, demonstration of national sovereignty, and development of specific mineral reserves. These more recent interests have involved the use of arctic marine shipping. However, it has been commercial interest in resource exploitation that has brought the full weight of modern science and technology to bear on the problems of arctic marine navigation. While the main objective has been to develop safe, dependable, year-round shipping in the Arctic, the research required to bring this about will undoubtedly advance our knowledge of the physical, biological, and social systems in the region.

The authors of this book have examined the convergence of these recent developments with the long-standing interest in arctic shipping. While they use specific examples for purposes of emphasis, their objective is to provide the reader with differing viewpoints and perceptions (sometimes quite contradictory) on the evolution of our approach to the use of arctic marine waters for shipping. The challenges posed by shipping proposals, however, are interconnected with the unique problems – scientific, technical, and socio-political – posed by the arctic setting.

Chapters 1–4 give different perspectives on three basic problems facing those who conduct science in the Arctic. The first problem is the lack of reliable data. As in any frontier area, little attention was given in the past to systematic collection and verification of information. Available records are often anecdotal or based on undefined sampling procedures. In technological development, world-wide experience with offshore exploration and development is not directly transferable to arctic conditions and must therefore, be interpreted by experts – who seldom agree.

The second problem is the tremendous variability that is characteristic of arctic systems. The environment – both physical (ice, currents, and water temperature) and biological (populations and migration routes) – exhibits such tremendous fluctuations through space and time that accurate sampling is difficult. Also, technology itself is changing so quickly; the scientist may not have adequate technical expertise to assess potential effects.

Third, the harshness of the environment, particularly of winter conditions, and the remoteness of the region make any data collection difficult and expensive. To overcome these problems, government has in the past combined its expertise with the logistical support of the private sector in joint research ventures. In spite of potential credibility problems, such team work will probably continue; neither partner alone can mount the effort required.

The Arctic also poses unique management problems. The ultimate determination of who will manage arctic shipping, and of the environmental tradeoffs involved, will be decided against a background of jurisdictional bargaining, political manoeuvring, and economic imperatives. Chapters 5–8 present the perspectives of the oil and gas industry, aboriginal peoples, and of the governments of the Northwest Territories and of Canada, respectively.

The basic issue is one of control. Whoever controls arctic shipping has a lever on the focus and pace of northern development and the attendant degree of risk to the northern environment. These management decisions are being faced at a time when the federal government is looking at the north as an opportunity to bolster its sagging economic fortunes but has also committed itself to improved environmental protection and management.

Some aboriginal groups have negotiated land claim settlements that give them some control over their economic and environmental affairs, and other settlements are pending. They see the offshore areas that they have historically used as part of their "lands," and they will undoubtedly try to exercise as much control as is allowable under the negotiated settlements.

The management issue is further confounded by the growing interest of the territorial governments in achieving provincial status and the pressures they are applying on the federal government to that end. During the past few years the territorial governments have been assuming greater responsibility for their own affairs. Major developments in arctic marine shipping could provide them both economic opportunity and a chance to further self-government.

This uncertainty over the roles and responsibilities of the major players in managing arctic developments, particularly vis-à-vis the environmental, social/economic, and cultural effects of arctic shipping, will take time to sort itself out. The authors of chapters 9 and 10 devote special attention to options in risk assessment, science policy, and ocean management for Canadian northern waters. However, one thing seems certain: the Road to Kingdom Come, as described by VanderZwaag in chapter 11, will be a rocky one indeed.

PART ONE

Science

CHAPTER ONE

In Pursuit of Knowledge:
Arctic Shipping and Marine Science
Cynthia Lamson

Historians of science, like jigsaw puzzle-makers, face two tasks. First, there is the need to examine the structure and organization of scientific enterprise. Puzzle-makers begin by assembling the straight-edged pieces to define the puzzle's frame or borders. The completed puzzle, a reconstructed picture, is achieved by examining contextual relationships between individual pieces and fitting together the pieces with complementary shapes and colours. Likewise, historians of science can look for "internal" links among ideas, personalities, and institutions to trace the development of an individual discipline or school of thought. They can also probe contextual or "external" factors influencing scientific achievement, recognizing often-critical exchanges among science, technology, and society.

The history of arctic marine science is a challenging topic for several reasons. First, because arctic marine science is not an identifiable discipline in its own right, it is necessary to survey activities spanning several centuries to "piece together" internal and external aspects of the enterprise. Second, there are two contrasting views about the origins of arctic science. The debate about what constitutes "science" and "scientific activity" revolves around questions of intent and method, and there is wide berth for interpretation in both. Formalists insist that early voyages of exploration – which made no systematic observations and measurements – belong to a pre-scientific era of arctic operations. Other scholars are inclined to label early polar navigators – with their keen observational and interpretive skills – the "founders" of arctic science.

The debate also raises theoretical questions about causation in history and science. Some would argue, like Thomas Carlyle, that history is shaped by singular, heroic, or charismatic individuals. Others maintain that ideas and the individuals who articulate them are products of society and time. Cogent arguments for both viewpoints could be made to describe and explain arctic marine science activity in general, and in Canada in particular. This essay presents a synoptic view of scientific activity in Canada's northern waters and endeavours to highlight key factors and personalities that have shaped the process and products of arctic marine science.

EARLY EXPLORATION AND OBSERVATION: 985–1939

E.F. Roots suggests that the history of arctic science is best characterized as the story of "very strong personalities and capable people with skill and vision, some of whom were not scientists themselves, but whose decisions and leadership profoundly influenced world knowledge and the pursuit of that knowledge" (1982a:374). He points to the accomplishments of Erik the Red and his wife, Thjodhild, who led to West Greenland, probably in AD 985, a colonizing expedition consisting of twenty-five ships and some 900 to 1,500 people. There were no navigational aids or shipping support services to assist the seafaring Vikings, and theirs was the largest arctic operation of any kind until construction in the 1950s of the 8046-km Distant Early Warning (DEW) line spanning across the Arctic from Alaska to Greenland.

Archaeological artifacts of Norse visitation and habitation have been recovered from sites as far north as Ellesmere Island and south to Maine, indicating that from earliest times the eastern Arctic was recognized as a potentially resource-rich, exploitable region. However, little is known about subsequent exploration activities until Martin Frobisher made three historic voyages to the southeastern coast of Baffin Island in 1576, 1577, and 1578. Frobisher was followed by a number of intrepid navigators, including John Davis, Henry Hudson, and William Baffin, who were charged with locating a northerly passage to the Orient for trading purposes. Although individual efforts failed to find a Northwest Passage, each successive voyage added to the growing body of knowledge about coastal geography, ocean currents, and ice regimes in arctic regions. For example, Baffin sailed north from Davis Strait to 77°25′N and observed and recorded an important open-water area, now known as the North Water polynya (Dunbar 1951:3–4).

The Europeans' failure to locate a navigable passage and their preoccupation with other parts of the world suppressed arctic exploration during much of the seventeenth and eighteenth centuries. Only the discovery of abundant whale resources in the Davis Strait and Baffin Bay region in the early nineteenth century refocused commercial attention on arctic waters. Smith (1931) credits Dutch and British whalers as the first dedicated ice observers in Baffin Bay/Davis Strait. Indeed, the exigencies of safety and commercial success were strong incentives to acquire a working knowledge of eastern arctic ice conditions. Records from the nineteenth-century whale fishery reveal that regional ice patterns were well known: "The description of the pack as 'middle ice' and 'best ice,' and the description of the movements of the pack we owe to Captain Marshall and Doctor O'Reilly (1818) of the whalers" (Smith 1931:4). Another whaler, William Scoresby, Jr, is cited by the British *Dictionary of National Biography* for inaugurating scientific activity in the Arctic. Scoresby's journals reveal his keen interest in observing, testing, and recording information about regional oceanographic conditions and about the characteristics, properties, and movement of ice (1820; rev. ed. 1969).

Debate about the origins of arctic science may be of interest to scholars; however, there is general consensus that political and economic circumstances were – and remain today – principal factors in promoting scientific enterprise. The end of the Napoleonic Wars in 1815 left Britain with a surplus of ships and men, and the Admiralty was anxious to keep them operational. In addition, trading ventures by the Hudson's Bay Company and others were continuing to expand in North America and along the western Greenland coast. This period of expanded trade coincided with reports from Scoresby and other whaling captains suggesting that climatic conditions in the eastern arctic might be moderating.

Thus, in 1818, the British Admiralty outfitted Sir John Ross with two ships, *Isabella* and *Alexander*, to search for a Northwest Passage. Ross, however, was also instructed to record his observations and conduct experiments to "contribute to the enhancement of science and knowledge," and Capt. Edward Sabine, Royal Artillery, was assigned to oversee sampling and data collection (Rice 1975:295). Ross terminated his voyage when he saw a high range of mountains closing off Lancaster Sound, and he returned to England. There his credibility was doubted by the British Admiralty, and in 1819 William Edward Parry was dispatched in *Hecla* and *Gripa* to pursue the quest for a Northwest Passage. Parry and his crews were able to penetrate as far west as Melville Island but were forced to overwinter at a site now known as Winter Harbour. Parry was extraordinarily successful in maintaining morale during the long polar winter, and all hands returned safely to England in 1820.

Subsequent nineteenth-century expeditions continued to push back the frontiers of arctic knowledge, but the disappearance of Sir John Franklin in 1845 was a milestone for arctic science. Some forty expeditions were launched to investigate the fate of Franklin and his crew, but his widow, Jane (Lady Franklin), is credited with insisting that each expedition also be equipped to undertake scientific work during its searching (Roots 1982a:383).

Over time, knowledge about ocean currents, ice drift, and ship resistance in ice has been acquired by systematic experimentation as well as by luck and ill-fate. For example, Capt. Henry Kellett's ship, *Resolute*, was abandoned at Beechey Island in Lancaster Sound in 1854; a year later she was discovered and salvaged in Davis Strait, thus becoming the first current indicator in the eastern Arctic. Other noteworthy episodes include the drift from Smith Sound to the Labrador coast by members of Charles Francis Hall's *Polaris* expedition (1871–3). The party became separated from the ship during supply-transfer operations and was eventually rescued by Capt. Bob Bartlett (Dunbar 1951:6–7). In 1880, Cmdr. G.W. DeLong's vessel, *Jeannette*, was crushed by pressure ice near Siberia, and fragments from the ill-fated vessel drifted across the polar ocean and were recovered near southern Greenland. Possibly the most famous drift "experiment" of all, however, was Fridtjof Nansen's deliberate besetment of *Fram*, a specially built ship designed to ride up on the ice to avoid being crushed. Nansen departed from Oslo in June 1893 and set fast to an ice floe at 78°50′N and 133°37′E. *Fram* drifted for three years with the pack ice and eventually broke free near Spitsbergen, thus demonstrating the existence of a transpolar drift stream (see Transche 1928; Nordenskjold and Mecking 1928).

Prior to designation of 1882 as an International Polar Year (IPY), scientific activity in arctic regions was primarily nationalistic, ad hoc, and conducted on an expedition-by-expedition basis. Karl Weyprecht, the visionary responsible for organizing the first IPY, argued persuasively to his peers that previous polar work was insufficiently rigorous for scientific needs: "Past arctic explorations were adventurous and of little value. They constitute an international steeplechase to the North Pole ... a system opposed to true scientific discoveries ... Immense sums have been spent and much hardship endured for the mere purpose of geographic and topographic observation, while strictly scientific observations have been given secondary status. The ultimate aim must lie higher than the mere sketching and naming in different languages of islands, bays and promonotories buried in ice" (Roots 1982a:384; Barr 1985). Weyprecht died before the IPY commenced but his place in the annals of arctic science is secure because of his relentless lobbying efforts to mobilize support for co-ordinated, co-operative, and international scientific research in polar regions.

In 1884, the Canadian Department of Marine and Fisheries and the Geological Survey of Canada sent the first government-sponsored expedition northward to Hudson Strait in *Neptune*, under the command of Lieut. Gordon. Observations on surface temperatures, tides, and ice were made to help assess the viability of a port facility to expedite shipment of western Canadian grain to markets in eastern Canada. Five stations were erected to record ice conditions daily in Hudson Strait, and a sixth post was set up on the Labrador coast; they were maintained until 1886. Subsequent Canadian expeditions, including the 1897 voyage of *Diana* under Capt. Wakeham, the 1903–4 cruise of *Neptune* led by A.P. Low, and Capt. Bernier's three expeditions in *Arctic* (1906, 1908–9, and 1910–11), also took surface temperature readings and recorded ice conditions and arctic flora and fauna, but no dedicated oceanographic studies were undertaken until the mid-twentieth century. In contrast, oceanographic work was an integral feature of Scandinavian, British, and American expeditions to the eastern Arctic and in the Arctic Basin from the 1860s onward. The Danes, for example, initiated a system among vessels trading between Denmark and Greenland to record surface water temperatures and to take routine water samples as early as 1897, and the Danish Meteorological Office began to keep annual records of ice conditions in 1897 (Dunbar 1951:11).

The collision of *Titanic* with an iceberg off the coast of Newfoundland in April 1912 was instrumental in changing public and governmental attitudes about ice and oceanographic research, yet Canada still lagged behind other polar nations in arctic marine science. The United States initiated seasonal ice patrols in 1912–13 in the southern Davis Strait/Labrador Sea/Grand Banks region ("iceberg alley"), and the International Ice Patrol was established in 1914. In 1928, two independent scientific expeditions studied oceanographic and ice conditions in the eastern arctic area. The US-sponsored *Marion* expedition surveyed oceanographic and ice conditions from Newfoundland north to the seventieth parallel, while the Danish *Godthaab* expedition conducted biological and oceanographic studies in Baffin Bay. Expeditions by *Marion* and subsequently by *General Greene* (1931, 1933, 1934, and 1935) represented the first effort to collect systematically temperature and salinity data, from some 2,000 surface and subsurface stations and thus facilitated the mapping of circulation and ice conditions in the region (see E.H. Smith 1931; E.H. Smith, Soule, and Mosby 1937). In 1931, the first major treatise on regional ice conditions was published, based on historical records collected by the British Meteorological Office, the US Meteorological Office, the Ice Patrol, the Mecking expedition (1906–7), and the *Marion* expedition (see E.H. Smith 1931:6).

Canadian oceanographic work in northern waters commenced in 1929 under the auspices of the Department of Marine and Fisheries. *Acadia* was sent to Hudson Strait and Hudson Bay, where subsurface temperature and density measurements were taken at various depths. However, the Hudson Bay Fisheries Expedition of 1930, with H.B. Hachey as chief scientist, is generally regarded as the first systematic effort to collect physical oceanographic data through measurement, sampling, and drift experiments (Hachey 1949; Dunbar 1951; Collin and Dunbar 1964; Campbell 1976). Most subsequent oceanographic research was concentrated in the Gulf of St Lawrence and Strait of Belle Isle areas and in the coastal waters of Newfoundland and Labrador.

WARTIME AND POST-WAR MARINE SCIENCE

The outbreak of the Second World War awakened concerns about Canadian sovereignty and continental security. In 1940, the RCMP vessel *St. Roch* was refitted and asked to transit the Northwest Passage under the command of Sgt. Henry Larsen. The journey from Vancouver to Halifax required twenty-eight months, but in 1944 *St. Roch* made a return transit to Vancouver in a record eighty-five days – the first vessel to complete the passage in a single season. (Only Roald Amundsen, in a forty-seven-ton herring boat, *Gjoa*, preceded *St. Roch* through the Passage, departing Christiania, Norway, in June 1903 and arriving at Nome, Alaska, in August 1906.)

The need for improved upper-air data to assist Allied military aviation stimulated co-operative US-Canadian development of a northern network of meteorological observation stations. The United States provided radiosonde equipment and personnel to eleven arctic stations, while Canadians operated two. Between 1947 and 1950, five jointly operated stations were established, at Eureka and Alert, on Ellesmere Island; Resolute, on Cornwallis Island; Isachsen, on Ellef Rignes Island; and Mould Bay, on Prince Patrick Island (see Thomson 1948; Rae 1951).

Submarine warfare activities and the need to define better the acoustical properties of underwater sound brought increased funding for co-operative, physical oceanographic research projects. Relationships established during the war continued afterward to benefit both the military and oceanographic research communities: important links had been formed between scientists at the Pacific Naval Laboratory, the US Naval Electronics Laboratory, the Pacific Oceanographic Group, the Institute of Oceanography, Canada's Defence Research Board, and the universities of British Columbia and Washington (Campbell 1976). Following the war, US Navy and Canadian vessels conducted hydrographic and oceanographic research in the Arctic Archipelago and eastern Arctic as part of Operation Nanook, Operation Packhorse, and Operation Frostbite (1946), Task Force 68 (1947), and Task Force 80 (1948) (see Dunbar 1951:19).

The decision to construct the Distant Early Warning (DEW) line also required scientific support activity, particularly hydrographic surveying, to assess the feasibility of transporting construction materials and station supplies by ship. Thus, in 1950, the Joint Canadian-US Beaufort

Sea expeditions were initiated. Canadian and American vessels took soundings in the western Arctic to delineate safe shipping routes and to define the approximate boundaries of the continental shelf. Ocean-ographic data were also collected. uss *Burton Island* occupied 42 offshore oceanographic stations in 1951, while the Canadian vessel *Cancolim II* took 47 stations in 1951 and 182 stations in 1952. In 1954, co-operative, bilateral oceanographic research in the Beaufort Sea region resumed, and hmcs *Labrador*, during part of her east-west voyage through the Northwest Passage, engaged in hydrographic work – the first time that Canadian oceanographers from both west and east coasts collaborated in joint arctic research (Campbell 1976:2161; Giovando and Herlinveaux 1981:9–13).

Four factors facilitated post-war development of Canadian arctic ma-rine science (and oceanography in particular): the founding of the Arctic Institute of North America (1945); creation of the Canadian Joint Com-mittee on Oceanography (April 1946); the work and influence of key ocean scientists such as H.B. Hachey, M.J. Dunbar, and J. Tuzo Wilson; and the construction, purchase, or commissioning of three arctic-capable ships – *Calanus* (1948), the first dedicated arctic research vessel of Can-ada's Fisheries.Research Board; *Blue Dolphin* (1948), operated under the auspices of the Arctic Institute of North America; and hmcs *Labrador* (1954).

In 1945, the Arctic Institute of North America (aina) was created in response to lobbying by American and Canadian scientists, government officials, and other public figures with arctic interests. On the Canadian side, a petition to Parliament for incorporation of the institute was submitted by Dr Charles Camsell, W.S. Rogers, G.R. Parkin, and Dr J. O'Neill. The objectives of the new institute were "to initiate, en-courage, support and advance by financial grants or otherwise the ob-jective study of arctic conditions and problems, including such as pertain to natural sciences, sciences generally, and communication" (9–10 George vi, c. 45). Offices were established in Montreal, at McGill University, and in New York, but subsequent relocations to Washington, dc, Ot-tawa, Anchorage, and Calgary reflected fluctuating financial support from host institutions. Fellows appointed to aina represented a wide cross-section of academic, government, and military interests.

Canada's Joint Committee on Oceanography (jco) (1946–59) had the task of co-ordinating intergovernmental arctic research and had mem-bers representing the Fisheries Research Board, the Royal Canadian Navy, and the National Research Council. Later, the Canadian Hy-drographic Service, the Meteorological Service, the federal Department of Transport, and the Defence Research Board were also represented. The jco's success in obtaining government support and research funding is generally attributed to its members' seniority (Campbell 1976:2159). In 1959, the JCO was reorganized as the Canadian Committee on Oceanography (CCO). The CCO was charged with an expanded research-co-ordinating role at the national level and gradually became the government's principal advisory body on international marine science (see Hachey 1961; Campbell 1976).

Members of the JCO and its successor, the CCO, organized into working groups based on areas of research expertise. The Working Group on Ice in Navigable Waters, created in 1957, was to develop ice-forecasting techniques; perform research on physical properties of ice, ice-air-water interactions, and patterns of ice movement and degeneration; and promote adoption of a standardized ice-reporting system.

Possibly the single most important ally of mid-century arctic research, however, was H.B. Hachey, chief oceanographer of Canada and oceanographer-in-charge of the Atlantic Oceanographic Group (AOG). Hachey had participated in the Canadian Fisheries Expedition of 1930 to Hudson Bay and was the first physical scientist to describe the area's general oceanographic and circulation patterns. Hachey's experience and seniority were major factors in governmental commitment of research dollars and ship time for northern marine science between 1947 and 1962 (Campbell 1976: 2157).

Other individuals also helped develop Canadian arctic science capability and train subsequent generations of marine scientists. M.J. Dunbar, for example, trained at Oxford prior to earning his PHD in zoology at McGill in 1941. From 1941 to 1946, he served as Canadian consul in Greenland, and he later returned to McGill with a faculty appointment. Dunbar organized a research program known as the Eastern Arctic Expedition (EAE) commencing in 1951; as a result of its activity, the Fisheries Research Board established an Arctic Unit at McGill. Invaluable experience gained in the EAE helped many of Dunbar's graduate students launch careers (notably E.H. Grainger, A.W. Mansfield, and I.A. McLaren). In 1965 the Arctic Unit, renamed the Arctic Biological Station (ABS), moved to new facilities at Ste Anne de Bellevue. The ABS was responsible for marine, freshwater, and anadromous fish studies for arctic and subarctic waters, as well as for marine mammal research north of sixty degrees (see Johnstone 1977).

Moira Dunbar deserves special mention for her pioneering sea-ice research for the Defence Research Board. She was the first woman to participate in summer research cruises aboard Canadian government ice-breakers and helped to develop sideways-looking radar for airborne ice reconnaissance. Dunbar also represented Canada at numerous international arctic science conferences.

The University of British Columbia established Canada's first ocean-

ography program in 1949. Later, after construction of the Institute of
Ocean Sciences in Sidney, British Columbia, research in arctic oceanography and marine science was undertaken by the Frozen Sea Research
Group (created in 1964). On the east coast, the Institute of Oceanography (later the Department of Oceanography) was established at Dalhousie University in 1959, and the Bedford Institute of Oceanography
opened in 1962.

The designation of vessels to undertake scientific work in the Arctic
marked another important milestone in Canadian arctic science. *Blue
Dolphin* was a heavy-planked wooden schooner, built in 1926 at Shel-
burne, Nova Scotia. She was refitted to meet the rigours of arctic
conditions and commissioned by the AINA to undertake scientific expeditions on the Labrador coast in 1949 and 1950. Although Canadians
did not participate directly, the Canadian government financially assisted the expeditions headed by Cmdr. David C. Nutt, US Naval Re-
serve, of Dartmouth College, in Hanover, New Hampshire (Nutt 1951).

In 1948, the forty-three-ton diesel ketch *Calanus* was designed for
arctic research and built at Mahone Bay, Nova Scotia. Between 1947
and 1956, *Calanus* was fully used for physical and biological research
in the eastern Arctic, including Foxe Basin, and in coastal waters off
Baffin Island. E.H. Grainger's research on board focused on data related
to plankton abundance, distribution and reproductive cycles (see Dunbar 1951; Collin and Dunbar 1964). HMCS *Labrador* was also built, ex-
pressly for marine scientific research, in 1954; however, after a brief
period of research activity, including hydrographic charting in the Beaufort Sea during the historic 1954 transit of the Northwest Passage, she
was transferred to the Canadian Coast Guard and deployed mainly for
other work (Dunbar 1981).

To summarize, Canadian arctic marine research throughout the 1950s
and into the 1960s was the relatively exclusive domain of a handful of
dedicated scientists with mutual interests. At civilian domestic levels,
information about arctic research activities was co-ordinated through
the intergovernmental Joint Oceanographic Committee, but the Arctic
did not become a priority area for intensive marine research until offshore hydrocarbon deposits were discovered during the 1970s. The
United States, in contrast, established at Point Barrow, Alaska, in 1947,
the Arctic Research Laboratory (later renamed the Naval Arctic Research Laboratory, NARL, under the Office of Naval Research), to conduct scientific work in arctic regions. NARL served as a support station
for scientific expeditions representing university, military, and commercial interests and was an unusual military institution in its encour-
agement of the exchange of ideas and information among users (see
Reed 1969).

IGY, "THE VISION," AND BEYOND (1957–69)

"Big Science" in arctic regions was first conducted during the first and second international polar years (1882–3 and 1932–3). Scientific research efforts were directed primarily toward meteorological and atmospheric investigations, but sea-ice studies were also conducted as part of overall program activities. Proposals for a third IPY were first tabled in 1950; however, it soon became apparent that a more comprehensive program of research was warranted. Thus, under the auspices of the International Council of Scientific Unions (ICSU), 1957–8 was designated International Geophysical Year (IGY), with sixty-seven nations participating in the largest-ever co-operative research undertaking. The oceanographic program involved more than one hundred ships making simultaneous observations on deepwater circulation, polar fronts, sea-level changes, and so on (see Hachey 1961; Rowley 1966).

The IGY coincided with John Diefenbaker's quest for a Conservative majority in Parliament. On 12 February 1958, he proclaimed his "vision" of a new Canada: "A Canada of the North! ... We intend to carry out the legislative program of Arctic research, to develop Arctic routes, to develop those vast hidden resources the last few years have revealed." The annual budget of the federal Northern Affairs Department nearly doubled, from $34 million in 1958 to $71.5 million in 1959, including substantial funding for northern airport, harbour, and icebreaker construction (Newman 1963: 217–23).

Additional evidence of the government's commitment to northern research was demonstrated when the Department of Mines and Technical Surveys began to organize the Polar Continental Shelf Project (PCSP) in 1958. The PCSP has been an ongoing investigation of the Canadian shelf area and surrounding waters since 1959, when eleven scientists were sent out to establish a 300-km survey line, take hydrographic and seismographic readings, and study gravity and magnetism. By 1985, more than 200 research teams had gathered voluminous data on sea-ice dynamics, glacier physics, paleoclimatology, the composition of the ocean floor, and many other topics (see Foster and Marino 1986). The PCSP continues to co-ordinate annual field studies and occasionally sponsors major projects such as the Arctic Ice Dynamics Joint Experiment (AIDJEX), a Canadian-US oceanographic program in the Beaufort Sea (1969–76), and the Canadian Exploratory Studies of the Alpha Ridge (CESAR), in 1983.

By the mid-1960s, however, the character of Canadian arctic science was beginning to change. Certainly expanded government interest in northern development helped gain needed research dollars and led to on-the-job training for succeeding generations of arctic scientists. But as government departments assumed greater responsibilities for science policy-making, bureaucrats joined and, in some cases, replaced scientists as members of science advisory boards. In addition, government bureaucrats also became more directly involved in project management. Thus, as J. Tuzo Wilson observed at a 1972 seminar, "Science and the North," bureaucratic control over arctic research was basically consistent with previous (national) initiatives to integrate policy-making with priority-setting. For example, the government created in 1916 its own National Research Council, and later myriad other science advisory bodies, such as the Defence Research Board, the Science Council of Canada, and the Ministry of State for Science and Technology.

Because Ottawa appointed representatives to boards set up to design science policy and set research priorities, the boundaries between political and scientific agendas could be obscured, often at the expense of sound and original science: "The plethora of bodies set up by the government to examine the supposedly poor state of government science have all failed to point out that the only thing wrong with Canadian science is the control government exerts over it ... As part of a trend towards bureaucratic lack of competition the government has annexed itself to most of the largest research projects in the Arctic. ... The Arctic is a graveyard of projects abandoned because they were ill-conceived and too hastily executed" (Wilson 1973:257).

MANHATTAN TO BENT HORN (1969–85)

The discovery of potentially exploitable hard mineral and offshore hydrocarbon resources in the Arctic ushered in a new era of northern scientific and technological research. Undoubtedly, the 1969 and 1970 transits through the Northwest Passage by Humble Oil Co.'s 155,000-ton, ice-strengthened supertanker, *Manhattan*, served notice on the Canadian government that inaction or ambivalence toward arctic resource development and marine environmental protection was tantamount to discrediting national claims of sovereignty over the waters of the Arctic Archipelago.

A series of legal, administrative, and policy measures was accordingly announced to demonstrate a federal commitment to the region, including enactment of the Arctic Waters Pollution Prevention Act (1970); proclamation of two national policies – the Northern Policy (1972) and the Oceans Policy (1973); issuance of Expanded Guidelines for Northern Pipelines (1972); creation of the Mackenzie Valley Pipeline Inquiry (1972–7); and support for the Beaufort Sea Project (1974 5). In 1977, the government also announced the "Make or Buy" policy, a somewhat radical initiative designed to reduce the amount of research and development (R&D) performed in government departments and to stimulate research in the industrial sector.

The Beaufort Sea Project (established 1973) was a joint government-industry research program to investigate the potential effects of exploratory drilling activity. Previous oceanographic work in the region had been carried out by the Frozen Sea Research Group (FSRG) of the Institute of Ocean Sciences (part of the then Ocean and Aquatic Affairs Division of Environment Canada). The FSRG was formed in 1964 to study oceanographic problems unique to ice-covered waters in response to government leasing for offshore exploration which commenced in 1963. After obtaining Beaufort exploration permits, the oil industry began a relatively intensive program of engineering research to determine how to cope with the arctic environment, and an estimated $2 million had been spent by 1973. In contrast, the government had failed to make comprehensive studies of the area, relying instead on data collected from a variety of earlier projects, such as AIDJEX and the Polar Continental Shelf Project.

However, probably because of concerns expressed by members of the FSRG, the Department of Environment intervened when the issue of approval for offshore drilling reached cabinet on 31 July 1973. This intervention led to a moratorium on drilling, pending the findings of an eighteen-month study to be jointly managed by industry and government. The Beaufort Sea Project was thus conceived. Although the project has been criticized from many quarters, the results helped alert scientists, industry, government officials, and northern residents to the fact that an offshore well blow-out could severely damage the arctic marine environment and that existing contingency plans and clean-up measures were inadequate (see Pimlott, Brown, and Sam 1976; Wadhams 1976; Milne n.d.).

The Beaufort Sea Project was Canada's first large-scale, multi-disciplinary environmental research effort. In total, forty-six technical reports were produced on a wide range of topics, from distribution and abundance of marine mammal populations, biological productivity, sea-ice morphology and oil-spill counter-measures through remote sensing technologies, environmental prediction systems, and socio-economic aspects of the use of renewable resource. Allen L. Milne, the federal government's project manager, assessed the effort as an inadequately designed two-year "crash" program. However, the data provided Dome, Esso, and Gulf with useful baseline information for their 1982 environmental-impact statement concerning hydrocarbon development in the Beaufort Sea–Mackenzie Delta region. Governments appear not to have learned as much from the experience, since they used the project as a model for subsequent pre-exploratory drilling reviews – for example, the Eastern Arctic Marine Environmental Studies (EAMES), the Offshore Labrador Biological Studies (OLABS), and the Baffin Island Oil Spill (BIOS) Project ("Beaufort Sea" 1982:20–1)

The decade 1975–85 witnessed a quantum expansion of arctic marine environmental and engineering research as industry requested exploratory drilling permits for offshore areas in the eastern, central, and western Arctic and in the Arctic Islands. Canadian marine science capacity was stretched to the limits, with government, industry, native and public interest groups competing to obtain sound data and expert advice (see Stirling and Calvert 1983). Requests for exploration permits and submission of proposals for non-renewable resource development made clear that decisions were being made on the basis of limited data and experience and minimal understanding of arctic ecosystems, the effects of industrial activity and vessel traffic on native harvesting patterns, and ship/ice interactions (see Roots 1981).

Ultimately, government uneasiness about scientific uncertainty resulted in two significant developments. First, the "need to know" emerged as a government priority, and several major R&D programs were created. For example, on 25 January 1981, Transport Canada, Fisheries and Oceans, and Environment Canada received cabinet approval for a five-year R&D program in arctic marine transportation, consisting of projects in regulatory-design, regulatory-operations, regulatory-general, operation of vessels, and government services.

Over the next few years, the federal government planned contributions to three arctic research programs. It put $26.4 million into the R&D program in arctic marine transport. The Environmental Studies Revolving Fund (ESRF) was established as part of the Canada Oil and Gas Act proclaimed in March 1982. The ministers of two federal departments – Energy, Mines and Resources and Indian Affairs and Northern Development (DIAND) – were given responsibilities for allocating monies from the fund for offshore-associated environmental and social science research. The seven-year Northern Oil and Gas Action Program (NOGAP) was approved in November 1984. NOGAP was established to support government "preparedness" for northern hydrocarbon production, with seven federal agencies designated as eligible recipients (Transport Canada, Fisheries and Oceans, Environment Canada, Energy, Mines and Resources, DIAND, National Museums, and Agriculture Canada). Initial commitments of $130 million were later reduced by 43 per cent, following overall federal expenditure reductions.

Second, the federal government re-examined science research and environmental assessment and review procedures. The need to consider ecological systems' behaviour in resource development and management designs was first articulated in the 1970s by academic scientists, who labelled their approach "adaptive environmental assessment and behaviour" (ESSA 1982). Subsequently, G.E. Beanlands and P.N. Duinker wrote *An Ecological Framework for Environmental Impact Assessment in Canada* (1983), a major critique of past assessment practice and a well-reasoned plea for new approaches. The federal government first applied the concepts and strategies described by Beanlands and Duinker in the 1983-4 Beaufort Environmental Monitoring Program (BEMP), which overlapped with review by the Federal Environmental Assessment and Review Office (FEARO) of the Beaufort Sea Hydrocarbon Production and Transportation proposal.

CURRENT STATUS

Governmental views about the purpose or appropriate objectives of arctic science seem to shift with disconcerting frequency, but the struggle between utilitarians and idealists has been a century-long battle. Karl Weyprecht, innovator of the first IPY, argued: "The Earth should be studied as a planet. National boundaries, and the North Pole itself, have no more or less significance than any other point on the planet, according to the opportunity they offer for the phenomena to be observed" (see Roots 1984:5).

In the first volume of *Arctic* – the journal of the Arctic Institute of North America (AINA) – A.L. Washburn reiterated Weyprecht's views: "Scientific problems are similar regardless of international boundaries, and the number of problems in the Arctic and Sub-Arctic that can be best solved by international cooperation is legion. In fact many of them can be solved only by international cooperation" (1948:4). Thirty-two years later, Washburn summarized advances in polar research, noting that in the IGY (1957–8) "polar research in the sense of modern science ... came of age." Scientists learned that many questions remained unanswered and thus offered new challenges to the scientific community. Washburn views the polar regions as natural "laboratories" of global significance but recognizes that science is linked to other goals such as national self-interest and economic opportunity (1980:643–52).

The "Arctic as global laboratory" view has persisted at least since Weyprecht's era but has not been shared by governments of nations with arctic interests. Thus the principal keepers of this vision have been universities and non-governmental organizations. For example, the American Geographical Society sponsored a symposium on arctic and antarctic research in 1928 to gain "support for well-qualified expeditions." The society also championed the view that "science, not adventure will be the ruling motive in future polar work" (Joerg 1928). The founding in 1945 of the AINA was another expression of support among US and Canadian scientists for co-operative research to further scientific understanding rather than nationalistic development goals.

A more utilitarian approach views the Arctic as a strategic resource depot. Proponents of this vision regard the Arctic principally as an engineering or technological challenge. Some of them express their impatience and frustration with research in terms of economics or national interest:

To the scientists I would say that sometimes scientists forget that, though their work is very important in itself, if it were not for engineers drawing out the good science and putting it to use for public wellbeing, the scientists wouldn"t get the funds. The engineers sometimes forget that … they have to know some of the underpinnings of science.

It is very important that, when we approach this problem of the Arctic, we do not widen the international participation to those who have no possible interest in the Arctic, that we keep it to those nations that surround the Arctic and have a real interest in it." (Gerwick 1985:11)

The utilitarians have generated more funding for arctic research than exponents of the global laboratory vision, because their research tends to have an applied focus. For example, testing the capacity of vessels and structures to withstand the pressures of ice, or designing systems to enhance meteorological and surveillance capacity, assists commercial development of arctic resources. Applied research proposals may be more "saleable" politically, because governments require continuing public support; results may be tangible and may appear more quickly in applied research than in long-term, basic research.

The preference for applied research may result also from the lack of scientific training or background among government decision-makers. A recent paper from the Science Council of Canada notes that only 5 per cent of members of Parliament have worked or trained in science, engineering, or medicine (Fish 1983:14).

In reviewing the relatively retarded development of science in Canada, Jarrell and Roos (1981) pose several questions relevant to the history of arctic science. "How has our branch plant economy aided or impeded, or been aided or impeded by, science and technology?" Ottawa's 1972 "Make or Buy" R&D policy may be informative in this instance. Arguably, the effort to stimulate private-sector growth, combined with promotion of regional development, exacted a price from the scientific community. Contracting-out produced an explosion of private consulting firms, and perhaps the consulting industry should be added to Jarrell and Roos's list of basic institutional forums for science (i.e. universities, scientific societies, and government). A leading example is

provided by A.E. Pallister, former vice-chairman of the Science Council
of Canada, who established a Calgary-based consulting firm in 1973.
Pallister has successfully linked private enterprise with arctic science
research by obtaining rights and contracts to publish data obtained in
publicly funded research undertakings, such as EAMES, OLABS, and the
Beaufort Environmental Impact Statement.

In summary, the tension between the global laboratory and strategic
depot perspectives has persisted through decades of scientific activity
in the Arctic. However, recognition of and response to such tensions
vary. B. Roberts, rapporteur at the Ditchley Park (UK) Arctic Ocean
Conference in 1971 was remarkably candid about exchanges of opinions
during the meetings: "One of the tensest and most protracted arguments
ever staged at a Ditchley Conference. It comprised several simultaneous
battles: between Canadians and others as to the whole regime of the
Arctic Ocean north of the American continent, between (or so it seemed
at times) business men on the one hand and scientists and lawyers on
the other, with political interests straddling the balance, between exploitation and defence of the environment" (1971:2).

Other responses to tension (or conflict) include withdrawal from
participation in projects and establishment of alternative research mechanisms or forums open to like-minded individuals. Conferences devoted
to arctic research and associated topics have proliferated since the early
1970s, but a comprehensive inventory of arctic-oriented meetings has
never been compiled. However, even a cursory survey of proceedings
of conferences reveals increasing specialization, suggesting that arctic
science is maturing. For example, the biennial POAC (Port and Ocean
Engineering under Arctic Conditions) conferences have evolved into
highly technical and professional forums since the meetings were initiated, for multi-disciplinary exchange, in 1971.

The conflict between utilitarians and advocates of basic science has
also produced a group of "Young Turks" – scientists willing to take
professional risks. A.R. Milne and B.D. Smiley of the Institute of Ocean
Sciences (IOS) were among the first marine scientists to express concerns
about the conduct of arctic science and the use of research findings in
decision-making. Both were involved in environmental research in the
Beaufort Sea and Lancaster Sound regions during the 1970s, and they
circulated their reservations about arctic industrial development in reports such as *Offshore Drilling in Lancaster Sound* (Milne and Smiley
1978) and *LNG Transport in Parry Channel* (Smiley and Milne 1979).
The effect of focusing public attention on scientific uncertainty was
immediate and profound. For example, several major environmental
and land use reviews were initiated, including the Lancaster Sound
Regional Study (1979–84) and environmental assessments of the Arctic

Pilot Project (1980) and the Beaufort Sea Hydrocarbon Development and Production proposal (1984).

Another example of the uneasy relationship between science and decision-making is explicitly stated in a technical report commissioned by the Arctic Research Directors' Committee (Department of Fisheries and Oceans): "Throughout the proceedings, discussion and reviews that have given rise to this document constant pressure has been applied to the Working Group to make it more concise and more definitive and, where facts run out, to attempt definitive judgements based on the best information available. Ultimately judgements have to be made in the preparation of environmental control legislation; it is therefore reasonable to expect that those more intimately concerned at the scientific level should make the primary scientific judgements" (L. Johnson 1983:11).

Burstyn (1968) has suggested that the American scientific community came of age in the period preceding the Civil War, when institutions employing scientists were founded largely to provide "R&D" services to the shipping industry. The history of Canadian arctic marine science illustrates a similar maturation sequence, where visions of industrial and commercial development have been tied largely to advances in arctic marine transportation technologies. But the coming of age of the Canadian marine scientific community – characterized by increasing specialization and technical competence – carries a potentially debilitating price tag, and D.J. Gamble warns that the application of Western applied science in northern settings is "crushing the very essence of what is vital to survival in northern societies. In the process we are losing what is vital to the expansion of our own notions of science" (1986:23).

Specialization and emphasis on reductionistic methods may be the hallmarks of mature science, but history distant and near warns us that faith (or arrogance?) in the capacity of science and technology can be mesmerizing. Arrogance contributed to the demise of Franklin's much-heralded expedition and partially explains *Titanic*'s encounter with an iceberg, *Ocean Ranger*'s meeting with a once-in-a-hundred-years wave, and the fate of the space shuttle *Challenger*, when lower-than-normal temperatures caused an inexpensive o-ring seal to fracture. Scientists and decision-makers alike have arrived at another threshold of development, where we pass from using science to push back frontiers to applying science, when appropriate, to expand our understanding and appreciation of the surrounding world. In many respects, science is like an ice-breaking vessel. The captains of science may benefit from modern instrumentation and advanced knowledge, but experience, skill, and sound judgment are the best security against unsuspected hazards.

CHAPTER TWO

The Physical Environment
Robert A. Lake

Historically, arctic marine shipping was confined to the resupply of arctic communities during the short summer shipping season. Resupply activities for the eastern Arctic were co-ordinated through Montreal, Quebec City, and Halifax; the Mackenzie River system was the major artery for supplying settlements in the western Arctic. Ship travel through the Northwest Passage was exceptional. With the arrival of hydrocarbon and mineral exploration, summer marine activities increased in the Beaufort Sea and around Melville Island, Strathcona Sound, and Little Cornwallis Island. This led to the development of new shipping routes and increased usage of the marine route around Alaska into the Beaufort Sea.

With technological advances in the design and capabilities of ice-breaking ships, the arctic operational season has been incrementally extended over the last decade. At present, mineral concentrates are shipped from Strathcona Sound and Little Cornwallis Island during the summer and autumn, and oil from the Bent Horn Project on Cameron Island is shipped during late August and early September. These seasonal activities may be precursors to year-round shipping if offshore development and production plans are approved. Although the fate of individual projects may hinge on economic and political factors, the prospects of expanding arctic shipping to a twelve-month season are quite high, given current concern about sovereignty and national security. Long-term implications of mastering the Northwest Passage on a routine basis could be sweeping: the length of the voyage from Tokyo to Le Havre would be approximately 40 per cent less than via the Panama Canal route.

The expansion of shipping and associated activities such as offshore drilling requires a comprehensive and improved environmental assessment capability. Environmental assessment involves examining potential effects of proposed activities on the environment, as well as effects of the environment on proposed activities. However, environmental assessment and the making of informed decisions about projects require understanding of the nature of the marine environment, including its characteristics and the physical processes involved.

This chapter describes the main features of the arctic marine physical environment – water levels, water properties, ocean currents, waves, and, most crucial for marine transport, sea ice – and concludes with a brief look at the quality of arctic marine data and their use in environmental assessment. The tides, currents, and waves in arctic waters are not extreme when measured against those encountered by shipping elsewhere in the world. The important feature is obviously ice, in the form of sea ice or icebergs. Ice constitutes the most serious and challenging constraint to arctic marine shipping, impeding movement and threatening ships and cargoes. The following description of the marine environment emphasizes ice and attempts to show how tides, ocean currents, and waves affect the growth, decay, distribution, and movement of ice. Examples of the management of sea ice in support of marine shipping are given, and the prediction of sea-ice and iceberg movement is discussed. Not all aspects of sea ice are negative, for ice helps reduce the magnitude of waves and currents in the Arctic and often provides a convenient surface for aircraft, drilling platforms, and research activities. The following description also illustrates the interdependent nature of the marine and atmospheric environment and discusses the problems associated with environmental assessment, including the gaps in both our understanding and the provision of data bases.

WATER LEVELS

Water levels vary in response to a number of causes, the astronomical tides being the most regular and predictable. Tides in the Canadian Arctic are primarily semi-diurnal or mixed and mainly semi-diurnal (Dohler 1966). That is to say there are two high and two low waters each lunar day (24.8 hours). The two main semi-diurnal constituents have periods near twelve hours that differ slightly in length; consequently the two signals go in and out of phase, creating spring tides (maximum water heights) and neap tides (minimum water heights) once every twenty-eight days. In Baffin Bay the average tidal progression proceeds northward along the western Greenland coast across northern Baffin Bay and then southward along the coast of Baffin Island, diminishing in amplitude until the tidal range reaches a height near 0.5 m along the central Baffin Island coast. Maximum tidal heights occur in northern Baffin Bay, where they attain values of 2.7 m. Maximum

heights of 1.2 m occur in Lancaster Sound near 85°w, north of the
Brodeur Peninsula. The Atlantic tide progresses westward through the
Arctic Archipelago from Baffin Bay and is superimposed on the smaller-
amplitude Arctic Ocean tide propagating eastward. Mean tidal ranges
attenuate from east to west through the Northwest Passage. Mean tidal
ranges are 1.9 m at Dundas Harbour (82.5°w), 1.3 m at Resolute Bay
(95°w), 1.0 m at Winter Harbour (110.7°w), and 0.4 m at Sachs Harbour, on the Arctic Ocean (125.3°w). Tides propagate through the
channels of the archipelago in the form of Kelvin waves. The wave
propagates forward with the coastline on the right. The amplitude is
greatest nearshore and decreases exponentially away from it.

In the Beaufort Sea, tides propagate from west to east along the coast
and counter-clockwise in Amundsen Gulf (Henry and Foreman 1977).
In the Beaufort Sea tidal ranges are relatively small, being 0.2 m at
Herschel Island (193°w), 0.3 m at Tuktoyaktuk (133°w), and a significantly higher 0.7 m at the entrance to Eskimo Lakes (131°w). In
Amundsen Gulf, ranges are similar: 0.3 m at Cape Perry (124.7°w),
0.4 m at Paulatuk (124°w) and Pearce Point (122.7°w).

Water levels vary in response to forces other than those associated
with the astronomical tides. The dynamic balance existing during geostrophic water flow requires a slope in the sea surface; consequently,
large-scale circulation patterns in the Arctic Ocean or Baffin Bay can
cause periodic changes in mean water level which appear to be sustained
over periods of weeks to months. Sea-level changes also occur in response to atmospheric pressure changes. At Tuktoyaktuk, water levels
have a marked seasonal variation, caused mainly by the discharge of
the Mackenzie River (Henry and Heaps 1976). The amplitude of the
semi-diurnal tidal constitutents also increases significantly from June to
October.

Strong winds associated with passing storms alter the mean water
level by piling water up against the coast or driving it offshore ("storm
surges"). In a review of monthly mean water levels in the Canadian
Arctic, Walker (1971) found that evidence in water level data for storm
surges was confined to the Beaufort Sea. The Beaufort Sea experiences
sea-level elevations well above normal whenever strong northwesterly
or westerly winds occur during the open-water season. Surges recorded
at Tuktoyaktuk are both positive and negative, with the latter occurring
irrespective of the degree of ice cover when the wind direction is offshore (Henry and Heaps 1976). Negative surges in the Beaufort Sea
have resulted in recorded water-level drops of one metre. Deep-draught
vessels operating in shallow waters in the Beaufort Sea are at risk during
negative storm surges, particularly when a significant decrease in water
level is accompanied by large-amplitude waves. Henry (1984) used the

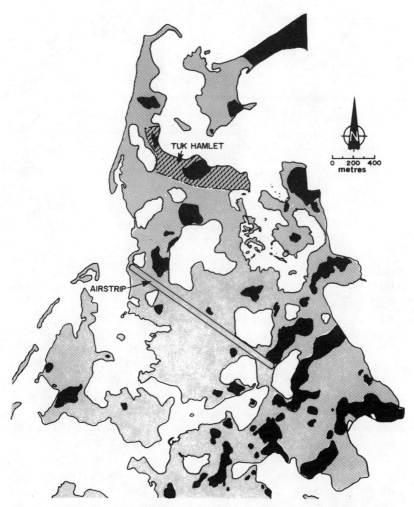

Figure 1. The delineation of estimated flood level due to a 100-year storm surge at Tuktoyaktuk. The dark shading indicates the area above flood level. After Henry, 1984.

available water-level data to determine the probability and consequence of storm surges on Tuktoyaktuk. Figure 1 shows his assessment of the area inundated by water during a hundred-year surge – i.e. a 0.01 probability of a surge of that magnitude occurring in a given year.

WATER PROPERTIES

The physical properties of sea water that are of most interest to us here are those related to density and heat content. The density of sea water

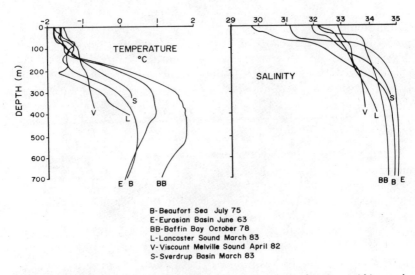

B- Beaufort Sea   July 75
E- Eurasian Basin June 63
BB- Baffin Bay October 78
L- Lancaster Sound March 83
V- Viscount Melville Sound April 82
S- Sverdrup Basin March 83

Figure 2. Profiles of temperature and salinity for representative locations within or adjacent to Canadian Arctic waters. Values may vary with the time of year. Adapted from Fissel et al, 1984.

depends on temperature, salinity, and pressure, and this section describes primarily the distribution of these properties in time and space in the surface waters of the Arctic. The vertical and horizontal gradients of these properties within the ocean sustain ocean currents, affect ocean mixing and the transport of oceanic heat, and are consequently related to the movement, growth, and decay of ice. Processes affecting ice are described in detail later in the text. Figure 2 shows typical profiles of temperature and salinity for regions of the Canadian Arctic.

The surface layer refers to the water extending from the surface to the depth, where there is a marked increase in density. This zone of increasing density is called the pycnocline. In arctic waters density depends mainly on salinity. A corresponding marked increase in temperature, called the thermocline, is usually located at the same depth as the pycnocline. The surface layer is subject to large seasonal changes caused by such factors as river runoff and sea-ice melt, large seasonal variations in solar radiation, turbulent mixing by moving ice and wind, and vertical convection, associated with brine rejection from growing sea ice.

In the Beaufort Sea the depth of the surface layer in late November was found to be from 9 m to 38 m (Melling 1983). The surface layer is vertically well mixed during winter by turbulence from wind and ice motion and vertical convection associated with growing sea ice. Salinity varied considerably – ranging at thirteen sites from 30.6 to 33.3

(dimensionless), at an average of 31.3 (practical salinity scale, 1978) – probably as a residual effect of Mackenzie River runoff and upwelling of saline water onto the shelf forced by summer winds (Melling 1983). Greater variability occurs during the summer. Herlinveaux, de Lange Boom and Wilton (1976) reported on summer conditions where surface salinities off the Mackenzie Delta and Tuktoyaktuk Peninsula were from 3 to 24 and temperatures were in the range of 8°c. Summer variability is also enhanced by wind-driven surface circulation.

The surface layer underlying the sea ice of the Arctic Archipelago is uniformly dense and at near-freezing temperatures. A synoptic survey of water properties within the archipelago in 1983 (Fissel, Knight, and Birch 1984) found that, along the northwest side of the Arctic Islands the surface mixed layer is relatively deep (22–50 m) and of lower salinity (31.5–32.0). Within the northern and western regions it is generally shallower (5–43 m) and more saline (31.5–32.3) and within 0.01°C of the freezing point. In the southern part, salinities are lowest (30.2–30.7), reflecting the influence of continental summer runoff. Within the central archipelago, surface waters are most saline (32.0–33.0) and warmer, because shallow sills enhance vertical mixing, which affects local ice growth. In the eastern archipelago, Lancaster Sound and Prince Regent Inlet were found to have surface layer depths from 11 m to 50 m and salinities of 32.3 to 32.8 and temperatures near freezing point. The channels of the archipelago experience cross-channel variations in surface layer, depth, and salinity, reflecting the adjustment of the density field to water flow.

Characteristics of water below the surface layer vary across the archipelago. To the west and south a temperature maximum may occur just below the surface layer, below which the temperature drops and then increases again, increasingly so below 150 m. Salinities increase steadily below the surface layer. In the north, both temperature and salinity increase steadily below the surface layer. In the centre, the water is mixed well by sills; consequently temperature and salinity may be uniform or increase a little with depth. In the east, the water column is mixed well, with a limited increase in temperature and salinity to depths near 150 to 200 m. Below that depth the increase of temperature and salinity marks the presence of Baffin Bay water.

OCEAN CURRENTS

Except in narrow channels such as Bellot Strait, Hell Gate, and Cardigan Strait, arctic currents do not significantly affect many aspects of marine transportation. However, they do transport and affect the growth of sea ice.

Ocean currents can be considered to result from the superposition of
two or more of the following influences: tidal forces; pressure gradient
forces related to changes in sea level; gravitational forces, resulting from
horizontal density gradients; wind stress directly on the water surface
or through an intervening ice sheet; and turbulent shearing (Reynold's
stresses).

Tidal currents occur as the result of tidal changes in water level.
Arctic tidal currents, like the tides, are mainly semi-diurnal. Tidal phases
and amplitudes may differ at opposite ends of a channel and may set
up long-channel currents in response to the resulting hydrostatic head.
Lake (1981) calculated that such a condition in Prince of Wales Strait
would give water-level differences between ends of the strait of 0.35
to 0.75 m, resulting in tidal currents between 50 and 100 cm/sec when
friction is taken into account.

The magnitude and direction of tidal currents vary with depth, because of frictional stress against the sea floor and landfast ice cover. The
major tidal current constituents, measured by four vertically displaced
current meters in Crozier Strait, for example, showed tidal current
speeds greatest away from the boundaries (Greisman and Lake 1978).
Vertical variability is also attributed to internal tide-related currents that
exist in a stratified water column and to sills that constrain flow below
sill depth.

Horizontal variations in tidal currents can occur across the channels
of the Arctic Archipelago. The tidal Kelvin wave propagates with maximum amplitude along the right-hand coast looking downstream. Lake
(1981) considered a shallow water wave propagating westward through
Barrow Strait. The calculated current speeds were 21 cm/sec along the
north shore and 16 cm/sec along the south shore. Boundary effects can
also produce horizontal variations in speed and direction, particularly
in shallow water and near headlands and spits.

Sea-level changes occur in response to atmospheric pressure gradients
and large-scale oceanic circulation. Aagaard (personal communication)
has suggested that a portion of the mean flow through the archipelago
may be sustained by sea-level differences between the north Pacific and
Atlantic oceans. Sea-level changes caused by Arctic Ocean circulation
can affect flow through the whole of the archipelago. Seasonal variations
in mean water flow of unknown origin were recorded in a year-long
current meter record from two sites near the sea floor in Crozier Strait
and are reproduced in Figure 3. Similar seasonal variations in flow,
recorded in Fram Strait between Greenland and Spitzbergen in the
North Atlantic, are consistent with long-term variation in flow related
to global oceanic circulation. Water flow will vary seasonally also in

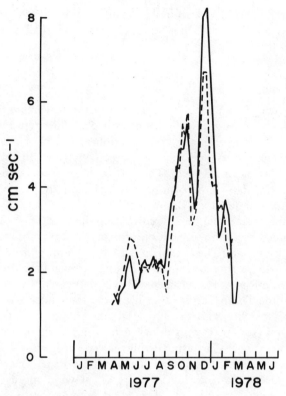

Figure 3. Seasonal variations in mean water flow at two locations in Crozier Strait. Values are averaged over successive seven day periods. Adapted from Greisman and Lake, 1978.

response to seasonal changes in density and, for shorter periods during summer, to wind stress.

During the period when sea ice is moving or entirely absent, wind stress can significantly affect water motion in a variety of ways. Wind-driven currents change water levels along coastlines where compensating flows are set in motion, affecting both surface and deeper layers. When the wind drops off, wind stress decreases and relaxation oscillations take place in bounded bodies of water, such as the channels of the archipelago, giving rise to standing waves and seiches. Surface currents set in motion by the wind continue to flow under their own momentum after the wind abates.

When water flow becomes turbulent, momentum is rapidly distributed by internal friction. These "Reynold stresses" convert energy from the oscillatory, or tidal flow to the mean flow. This component of mean flow, generally considered negligibly small, may become significant in narrow passages, such as Bellot Strait, or inlets with shallow sills, such as Bridport Inlet.

Beaufort Sea

The movement of the Beaufort Gyre dominates movement of surface water in the offshore zone of the Beaufort Sea. The gyre moves in a clockwise direction, with an average peripheral speed of 4 cm/sec, in response to wind stress on the ice associated with the mean atmospheric pressure field (Coachman 1969). The gyre does not extend much further south than the edge of the continental shelf, near the 200-m contour (Herlinveaux and de Lange Boom 1975).

During the open-water period, currents in the southern Beaufort Sea are determined by the superposition of many factors, resulting in variable surface currents. The dominant influences on mean drift are winds associated with passing atmospheric disturbances and the discharge of the Mackenzie River. The Mackenzie River plume tends to travel eastward along the Tuktoyaktuk Peninsula under the influence of the Coriolis force. Variability is increased by local winds and small-scale systems. MacNeill and Garrett (1975) studied surface currents during the open-water period in 1974–5, by using drifting surface drogues. Observations indicate that surface currents are chiefly wind-driven. The northwest wind had the greatest effect, especially when wind speeds exceeded 10 m/sec for extended periods. This wind produced strong southeast currents offshore and a northeasterly long-shore current along the coast. Herlinveaux and de Lange Boom (1975) report current speeds exceeding 60 cm/sec under these conditions. Strong winds from the east or northeast produce northwest currents offshore and long-shore westerly currents next to the coast, with observed current speeds over 50 cm/sec. During periods of calm or light winds, large-scale oceanic eddies and local winds resulted in a more variable current field. Overall, the predominant current direction is northeast, reflecting the dominant effect of northwest winds and the Coriolis force on the Mackenzie River plume. An exception is the consistent northwest long-shore current west of the Mackenzie Delta, which is determined by the bottom topography. The predominant current pattern in the Beaufort Sea is illustrated in Figure 4.

Arctic Archipelago

The net circulation through the Arctic Archipelago is considered to be from the Arctic Ocean toward the south and east into Baffin Bay.

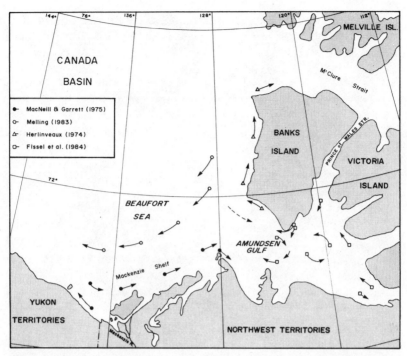

Figure 4. General surface current patterns in the southeastern Beaufort Sea and Admunsen Gulf.

Current measurements in the archipelago have provided data on water movements during the period from January to June (Greisman and Lake 1978; Peck 1978; Van Ieperen 1981). In addition, a large number of recent current measurements beneath the ice have been taken in the Northwest Passage. The results are not yet published but have been used in this description. Other studies have used drifters (Fissel and Marko 1978), sea-ice trajectories derived from satellite data (Marko 1977), and land-based radar (MacNeill, de Lange Boom, and Ramsden 1978) to examine the circulation of surface water during summer.

In general, tidal currents within narrow channels, such as Prince of Wales Strait, or adjacent to coastlines tend to be rectilinear, paralleling the shoreline and reversing flow during each half of the tidal cycle. Away from the coast of the wider channels, such as Viscount Melville Sound, tidal currents tend to be more circular, with the direction of flow rotating through 360 degrees on each cycle. Current patterns in the western part of the Northwest Passage are depicted in Figure 5 for the ice-covered ocean during spring.

Surface flow in Barrow Strait and Lancaster Sound is illustrated in

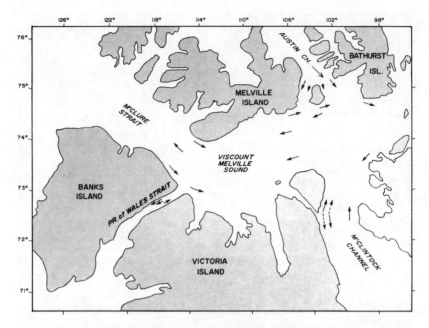

Figure 5. General surface current patterns in Viscount Melville Sound and adjacent channels. Based on Greisman and Lake, 1978, and recent unpublished data.

Figure 6. The information on surface flow patterns has been obtained
by tracking drifting buoys from July to November 1977 (Fissel and
Marko 1978); by observing ice drift from shore-based radar at the east
end of Griffith Island, south of Cornwallis Island, from mid-July to
mid-August 1977 (MacNeill, de Lange Boom, and Ramsden 1978); and
from moored current-meter data from both the spring (ice-covered)
and summer periods, from a variety of sources.

Current patterns determined during the open-water period are most
variable, because of the strong influence of local winds. Currents determined by all measurements, however, show some consistent flow
patterns. Currents flow eastward along the south side of Barrow Strait
and Lancaster Sound. A westward, though variable flow, is located
along the south coast of Devon Island. Currents flowing southward in
Baffin Bay, off the east coast of Devon Island, turn westward and flow
for 35 to 75 km into Lancaster Sound, where they turn and flow southward across the sound. In the south side of the sound this current may
turn eastward to flow back into Baffin Bay or may divide off the north
coast of Baffin Island, with one branch flowing eastward and the other
circulating clockwise to the west before returning to the centre of the
sound.

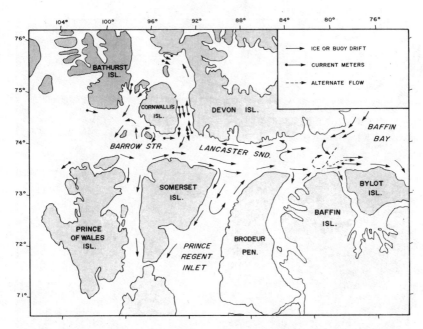

Figure 6. General surface current patterns in Barrow Strait, Lancaster Sound and adjacent channels. Based on Fissel and Marko, 1978; Greisman and Lake, 1978; MacNeill et al, 1978; Fissel et al, 1982; Topham et al, 1983 and unpublished data.

Baffin Bay

The general pattern of surface currents in Baffin Bay is shown in Figure 7. Two main currents dominate surface circulation – the northward-flowing West Greenland Current and the southward-flowing Canadian (or Baffin) Current, off the east coast of Devon and Baffin islands. Water flows into Baffin Bay from the north through Smith Sound and Nares Strait (Sadler 1976) and from the west through Jones Sound (Ito 1981) and Lancaster Sound. The current flow off eastern Devon Island and Baffin Bay has been described by Fissel, Lemon, and Birch (1982). From 8 to 35 km off the east coast of Devon Island the southward flow reaches speeds of 58 cm/sec. Off the east coast of Bylot Island the current continues southward at speeds between 30 and 75 cm/sec. Further south, off Baffin Island, currents on the continental slope, as measured by drifters, ranged from 12 to 24 cm/sec, compared to 24 to 40 cm/sec over the slope.

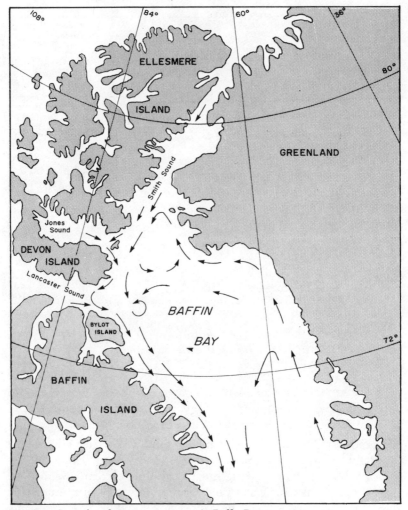

Figure 7. General surface current patterns in Baffin Bay.

WAVES

In the Arctic, locally generated wind waves occur in areas of open water, while ocean swell, a longer-period wave, can be generated elsewhere and occur in both open water and in marginal ice zones. The magnitude of waves depends on the fetch – the distance over which the wave-generating wind blows. Other factors aside, sea ice aids vessels' progress by impeding formation and propagation of waves. The speed and direction of waves will vary with water depth, bottom topography, and

Table 1
Percentage Frequency of Wave Heights

AUGUST						
Wave height (feet)	0–2	3–6	7–9	10–12	13–19	20+
Beaufort Sea	54.9	20.7	16.4	6.1	1.9	0.1
Amundsen Gulf	75.6	22.5	2.0	–	–	–
Visc. Melville Sd.	87.5	–	12.5	–	–	–
Barrow Strait	90.0	10.1	–	–	–	–
Lancaster Sound	67.9	27.3	2.8	1.6	0.4	–
N. Baffin Bay	68.4	26.5	1.7	–	3.4	–
SEPTEMBER						
Beaufort Sea	20.5	31.1	28.8	8.3	9.8	1.5
Amundsen Gulf	51.6	37.1	11.3	–	–	–
Visc. Melville Sd.	80.9	19.0	–	–	–	–
Barrow Strait	63.5	26.1	4.3	6.1	–	–
Lancaster Sound	50.5	36.0	9.8	3.2	0.5	–
N. Baffin Bay	59.1	30.8	6.0	2.7	1.0	0.3

After Parker and Alexander 1983.

wave-length. Although significant waves may be generated in larger polynya such as North Water during the winter, waves are primarily a summer phenomenon. The period of significant wave heights occurs in the Beaufort Sea during August and September and in Baffin Bay from July through October.

Parker and Alexander (1983) have summarized the frequency of wave heights for zones of interest to arctic marine transportation. Data for August and September are reproduced in Table 1 for selected locations. The smaller wave heights in the Arctic Archipelago are a consequence of the limited fetch resulting from the islands and the presence of sea ice.

The high inter-annual variability of ice cover and atmospheric conditions makes determination of valid wave climate statistics difficult. A given set of wave data will reflect the conditions prevalent during the data collection period, and so records representing several years are generally required. Wave statistics are often generated from knowledge of the meteorological conditions and the extent of ice cover (i.e. the fetch); however, meteorological data are sparse to non-existent offshore, and the understanding of wave development and propagation in partially ice-covered waters is poor.

Hodgins (1983) reviewed the extreme wave conditions in the Beaufort Sea based on the best available analyses and estimated that deep-water (greater than 80-m depth) significant wave heights were from 6 to 8 m and from 7 to 9 m for fifty- and one-hundred-year return periods, respectively. Wave heights over the shallower continental shelf may vary somewhat, depending on wave-length. Lachapelle (1981), using available wind data for central Lancaster Sound, computed annual significant wave heights of 9.4 and 10.8 m for fifty- and one-hundred-year return periods. In Baffin Bay, off the entrance to Lancaster Sound (77°w), significant wave height for the 100-year return period was 12.9 m.

SEA ICE

The feature of the arctic marine environment that most affects marine transport is sea ice, which will be considered under six general headings: introduction, significant ice features, sea-ice growth, sea-ice decay, ice break-up, and modelling ice behaviour.

Introduction

Definition: Sea ice refers to any type of ice originating from the freezing of sea water. The freezing point of sea water varies with both water salinity and pressure or water depth. When the temperature of the sea surface falls slightly below its freezing point, ice crystallization occurs in the form of ice needles. In a calm sea, ice needles grow and change to plate crystals. These crystals join, forming a continuous matrix of ice crystals on the water surface. Sea ice grows to a compact aggregate of roughly columnar, lamellated crystals, 0.05 to 0.1 cm wide. Crystals with vertical lamellae can conduct more heat and consequently grow more rapidly at the expense of surrounding crystals. The salt contained in sea water is excluded from the ice-crystal matrix, and makes brine that is saltier than the original sea water. Most of the new dense brine drains into the underlying sea water; the remainder, along with air bubbles, is trapped in pockets within the ice, giving the ice its unique properties.

Classification: Sea ice is classified according to an ice nomenclature approved by the World Meteorological Organization. This nomenclature is used by the Atmospheric Environment Service (AES), which observes, records, and reports ice conditions in Canada (Department of Transport 1965). The nine most commonly referred-to ice types, as defined in the ice-observing manual, are frazil ice: fine ice spicules or plates of ice suspended in water; grease ice: ice crystals coagulated to form a soupy layer on the water surface; nilas: a thin elastic coat of ice, bending on waves and thrusting in a pattern of overlocking fingers; pancake ice: circular pieces of ice with raised rims about 30 to 250 cm in diameter; grey ice: young ice 5 to 10 cm thick; grey-white ice: young ice 15 to 30 cm thick; first-year ice: sea ice of not more than one year's growth;

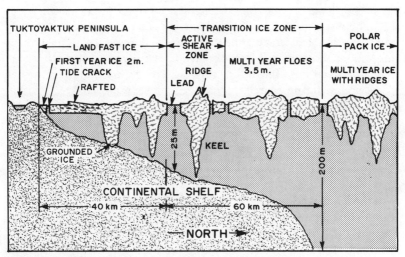

Figure 8. The major features of ice zones in the Beaufort Sea.

second-year ice: old ice that has survived only one summer's melt; multi-year ice: old ice up to 2.7 m thick, which has survived at least two summers' melt.

Under appropriate conditions, visible and near-infrared imagery from satellites such as ERTS and Landsat can distinguish between grey, grey-white, and older ice forms (Barnes and Bowley 1974).

Distribution: In the Beaufort Sea, the winter ice cover is divided into three zones, as illustrated in Figure 8. The nearshore Landfast Ice Zone extends for approximately 40 km off the Tuktoyaktuk Peninsula, corresponding roughly to the 25-m-depth contour. Landfast ice grows out from shore, beginning in late September to early October, until it reaches the normal maximum extent. This zone is comprised of first-year ice, but may incorporate some multi-year ice, rafted ice, and pressure ridges. Pressure ridges tend to run parallel to the shoreline and become grounded in the shallow water. Annual sea ice will grow to the sea floor to a depth of 2 m, to create bottom-fast ice. Tidal excursions are accommodated by tide cracks at the seaward edge of the bottom-fast ice.

The Transition Ice Zone exists between the Landfast Ice Zone and the polar pack ice – roughly from 40 km to 100 km offshore, where water depths are near 200 m. The Transition Ice Zone incorporates an active shear zone between the landfast ice and the westward-drifting polar pack, where flaw leads, oriented parallel to shore, open and close under the influence of the wind. This shear zone and a major flaw lead

extend northward, separating the landfast ice of the Arctic Archipelago
from the polar pack. The Transition Ice Zone is composed mainly of
multi-year ice. Ice deformation and ridge building occur actively here
throughout the winter. Mean drift tends to be westward, with maximum speeds near 11 km/day during September and minimum speeds
of 3 km/day during March.

The Polar Pack Ice Zone, located about 100 km offshore, is a permanent feature of the Beaufort Sea. It consists of multi-year ice and
ridges which move westward a few kilometres per day. The motion
of the polar pack occurs as part of the clockwise rotation of the Beaufort
Gyre, driven by the atmospheric high pressure normally located over
the sea.

In the Beaufort Sea, spring break-up becomes evident from late April
to mid-May, as the Bathurst Polynya and the flaw leads expand, increasing the extent of open water beyond the Landfast Ice Zone. The
extent of open water is less to the west near the Alaska coast, because
of the convergence of ice off Point Barrow. The landfast ice begins to
disintegrate in June, as the effect of the Mackenzie River spring runoff
is felt. The break-up of landfast ice is highly variable, reflecting the
inconsistency of offshore winds. Vessels overwintering in coastal waters
can commence operations only when they are able to clear the landfast
ice, either after natural disintegration or by using ice-breaking vessels.
Landfast ice may clear from coastal waters as early as late June, and
normally by mid-July; in very occasional years it may not completely
clear at all. The maximum extent of open water in the Beaufort Sea is
reached in late August or early September. The median position of ice
edges for late August in good, fair, and poor years is shown in Figure
9. During summer, vessels operating in the Beaufort Sea are less con-
strained by ice than vessels to the west, off Alaska, and to the east, in
the Northwest Passage. Ice break-up in Amundsen Gulf may occur as
early as April, as the Bathurst Polynya–flaw lead system extends into
the western gulf. In general, however, break-up begins in late June, as
part of the normal ice-disintegration process.

Within the Arctic Archipelago, optimum ice conditions normally
occur in early September, as illustrated in Figure 10. Normally, open
water exists in Amundsen Gulf, Barrow Strait, Lancaster Sound, and
adjacent channels to the south. In M'Clure Strait, which separates Viscount Melville Sound from the Beaufort Sea, ice concentrations are in
the range of 4/10ths to 9/10ths, inhibiting normal vessel traffic, which
is routed south, through Prince of Wales Strait. During the worst ice
years, open water is found only in portions of Amundsen Gulf and in
Lancaster Sound, the remaining area being generally covered in 6/10ths
to 10/10ths ice.

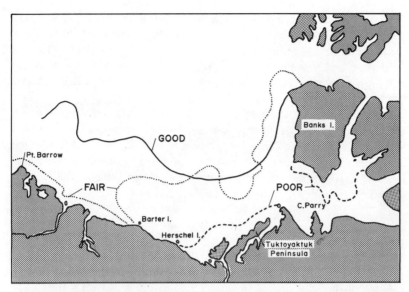

Figure 9. Median position of ice edges in the Beaufort Sea for late August in good, fair and poor years, 1961–1972. After Milne, undated.

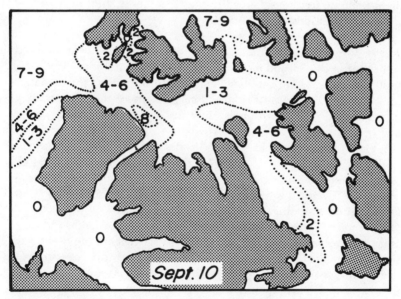

Figure 10. Median minimum ice concentration in the Arctic Archipelago. Values are in tenths of old ice concentration. 2 = 1/10 − 3/10. 8 = 7/10 − 9/10. Adapted from Parker and Alexander, 1983, following Markham, 1981.

In the northern archipelago, sea ice – mainly multi-year – remains year round. During the summer, ice is less consolidated, because of disintegration, and it becomes mobile. Ice is again landfast as freeze-up occurs, and subsequent motion is generally limited to a few metres, although occasional larger excursions have been noted at drill sites in the Sverdrup Basin.

In the eastern Arctic, northern Baffin Bay and Lancaster Sound are free of ice well into October, except in the worst years, when some ice floes may exist in the centre of the bay and off Baffin Island. The ice edge in Lancaster Sound does not become fixed until mid-winter, at a location that can vary from Barrow Strait in the west to the eastern end of the sound.

Significant Ice Features

Marine activities in the Canadian Arctic have been supported and advanced by significant development of technology to operate in ice-covered waters. Drilling platforms and ice-breaking vessels in the Beaufort Sea are examples; they can operate in first-year ice without difficulty. Of particular concern to marine operations, however, are such significant ice features as pressure ridges, icebergs, and multi-year ice floes and ice islands.

Ridges: Pressure ridges result from deformation of adjacent ice sheets which are in contact and moving with respect to each other. If the local motion is compressive, interfingering occurs. The overriding ice ploughs ice blocks and snow into ridges. The portions of the pressure ridge above and below the surrounding undisturbed ice are called the sail and keel, respectively. A cross-section of a pressure ridge is shown in Figure 11. The broken ice blocks making up the ridge are composed usually of thinner, first-year ice, 50 cm or less thick (Weeks, Kovacs, and Hibler 1971). Ice sheets undergoing shear can also result in ridges, typically straight and vertically walled on one side. Weeks, Kovacs, and Hibler (1971) report that ridge sails are rarely more than 5 m high. Kovacs (1983) reports extreme values for sail heights and keel depths of 12.8 m and 50 m, respectively, and a mean keel-to-sail height ratio of 1:3.3 for the Beaufort and Chuckchi seas. Dickins and Wetzel (1981) report larger and more variable results in the northern Arctic Archipelago, with a mean keel-to-sail height ratio of 1:5.6 (+/ − 2.2), based on twenty ridge cross-sections. It was theorized that this larger ratio may be attributed to the colder mean temperatures of air and water. If ridges survive the melt season, the inter-block voids within the keel become filled with refrozen melt water, and the strength of the ridge increases. Ridges that

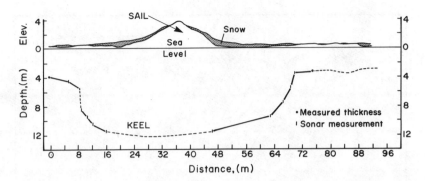

Figure 11. A cross-section of a multi-year pressure ridge. Adapted from Weeks et al, 1971.

survive into a second winter ("multi-year ridges") are significant obstacles to ice-breaking vessels and offshore structures.

Along the western edge of the Arctic Archipelago, the ridging of
multi-year ice sheets creates hummock fields of ridges and crushed ice,
which may exceed 20 m in height and 60 m in depth (Dome, Gulf, and
Esso 1982). As with multi-year ridges, refreezing of melt-water in
interstitial spaces increases strength, so that multi-year hummock fields,
which may be a few hundred metres across, are formidable obstacles
to ships and offshore structures.

Icebergs: Icebergs are fresh-water ice masses that break off, or "calve,"
from tide-water glaciers in Greenland and, to a much lesser extent,
Ellesmere and Devon islands. They are common in Baffin Bay, Jones
and Lancaster sounds, and the adjacent fjords and sounds of the eastern
Arctic Archipelago but are rare in the Beaufort Sea and the western
archipelago.

Icebergs are massive, up to several square kilometres in extent. More
commonly they extend over tens to several hundreds of square metres.
Smaller pieces of glacial ice – from 30 to 90 sq m in area (bergy bits)
and from 6 to 30 sq m (growlers) – also pose significant hazards to
vessels. Indeed, growlers cannot always be detected by ship's radar
when large waves exist.

Iceberg observations in Baffin Bay and Davis Strait indicate that
heavy concentrations usually decrease exponentially with distance from
the west coast of Greenland and that, although patterns of concentrations are generally consistent, there is considerable inter-annual variability in total numbers of icebergs (APP 1981b).

Most icebergs originate on the west coast of Greenland and drift in
a northwesterly direction in the Greenland Current. They subsequently

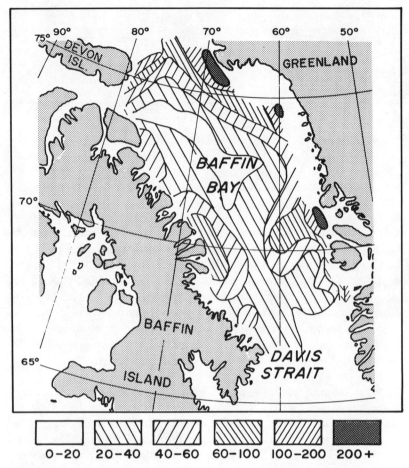

Figure 12. Composite average iceberg concentration in Baffin Bay and Davis Strait. July–October 1948, 1949, 1964–69. Values are for iceberg sightings per two degree square. After Arctic Pilot Project, 1981.

circulate counter-clockwise over the north end of Baffin Bay, returning south on the Canadian or Baffin Current, along the east coast of Baffin Island. Icebergs circulate into the eastern end of Lancaster Sound, usually returning to Baffin Bay along the south side of the sound. The composite average iceberg concentration in Baffin Bay over eight years is shown in Figure 12. Dietrich, Zorne, and Hasle (1979) present a scatter diagram of iceberg height and draught based on a study of 113 icebergs off the west coast of Greenland. Average height and draught computed from the diagram are 33 m and 83 m, respectively. A great deal of scatter exists in the ratio of height to draught: the extreme height found was 80 m, and the extreme draught (a different iceberg) was 187 m. Iceberg drift is determined primarily by the forces of ocean currents and the wind.

Thick Floes and Ice Islands: Sea-ice floes of exceptional thickness have been reported in the Arctic Ocean and in the western channels of the Arctic Archipelago (Serson 1972). Thermodynamic models (Maykut and Untersteiner 1969) gave equilibrium thicknesses of varying values, depending on the parameters chosen. The maximum ice thickness was found to be about 7 m, with an insulating snow cover 120 cm deep. Thick sea-ice floes, however, have been reported, with depths of 10 to 12 m. Walker and Wadhams (1979) have shown that sea ice might reach thicknesses near 20 m in the present climate in locations with little or no oceanic heat flux and annual snowfall of about one metre. Such conditions might exist at high-latitude ice shelves, such as those in northern Ellesmere Island, Svalbard, and northeastern Greenland.

The ice shelves of northern Ellesmere Island originate from glacier tongues nearly 50 m thick. The floating glacier tongues occasionally break off to form ice islands (Hattersley-Smith and Serson 1970). Ice islands circulate for years with the Beaufort Gyre, sometimes drifting eastward past Greenland, where they are swept south with the prevailing currents to melt in the North Atlantic. Initially ice islands have an areal extent of tens of square kilometres. However, during their lifetime they break into increasingly smaller sizes and become thinner through ablation. Fortunately, thick sea-ice floes and ice islands are rare, and the chance of vessels and offshore structures encountering them is not great.

Sea-Ice Growth

The formation of sea ice is a complex process and varies in different arctic regions, as we shall see immediately. It can be seriously inhibited, either naturally, resulting in polynyas, or artificially, by human intervention, as we shall see later in this subsection.

Formation: Sea-ice formation begins when the sea's surface temperature drops to freezing point, usually between 0 and −1.6°C, depending on surface-water salinity. To reach this state, the oceanic surface layer must lose heat obtained directly during the brief arctic summer through solar radiation, runoff from land, and rainfall. Heat loss to the atmosphere increases rapidly when the mean daily temperature falls consistently below 0°C, a condition that varies with solar radiation (that is, latitude) and patterns of atmospheric circulation. This condition is generally prevalent in the Arctic by mid-September. The time required for surface water to cool to its freezing point depends on sub-zero air temperatures and distribution of heat content in the oceanic surface layer. The distribution of heat depends critically on late summer storms, which extract heat and mix remaining heat more uniformly in the water column and to greater depth. Freezing is thus first evident in shallow bays, where heat is extracted quickly from water of limited depth.

Established sea ice grows directly in proportion to the degree to which latent heat associated with the formation of new ice transfers heat from the ice to the atmosphere. The latter heat flux will depend on the vertical thermal gradient (or temperature change per unit of ice thickness) and the thermal conductivity of sea ice. The temperature difference between the atmosphere and the ocean determines the thermal gradient. As the temperature of the ocean varies only slightly once it reaches its freezing point, the air temperature, with its variations from o to $-50°c$, controls the temperature gradient through the ice sheet. Simple models of ice growth relate ice thickness to accumulated "freezing degree days" (Bilello 1961). The thermal conductivity of sea ice depends on both the salinity and temperature of the ice (Schwerdtferger 1963) and has values near 4.3×10^{-3} cal/cm sec c°.

Heat sources contributing to the heat flux are latent heat associated with ice growth, heat loss associated with ice cooling (Untersteiner 1964a), and the heat removed from sea water in contact with the ice bottom in order to bring it to its freezing point (sensible heat). These and other processes related to ice growth and decay are shown in Figure 13.

First-year sea ice will typically be 2 m thick by the time ice growth ceases, usually during late April or May in the Canadian Arctic. Summer melt will thin the ice, but if it survives until the onset of freezing in September it will grow further, beyond the original 2 m. Thicker sea ice has a smaller temperature gradient, and ice growth progressively shows. Sea ice that undergoes several annual cycles of growth (perennial sea ice) reaches a maximum ice thickness, and no net ice growth occurs on an annual basis.

The heat flux through sea ice and consequent ice growth are inhibited greatly by snow cover, which insulates the ice from the cold air. The thermal conductivity of snow is significantly less than sea ice and is strongly dependent on snow density (Abels 1892). The average density of newly fallen snow at seven arctic weather stations was 0.09 gm/cm³ (Walker and Lake 1973), while at the onset of melting the snow density increases to 0.45 gm/cm³ (Maykut and Untersteiner 1969), as meltwater penetrates the snow and is refrozen. The nature of the mean snow cover, and consequently its effect on sea-ice growth, vary over the

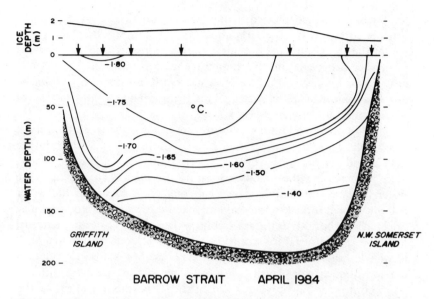

Figure 13. Water temperature structure and ice thickness across Barrow Strait. Arrows indicate measurement sites.

Canadian Arctic. Walker and Lake (1973), using AES data, found that annual precipitation normals for 1941–70 were near 15 cm along the northern continental coast and northern Baffin Island and tended to decrease to 10 cm along the northwestern coast of the Arctic Archipelago. For the same period, the proportion of normal annual precipitation falling as snow increased from 50 per cent along the northern continental coast to 80 per cent along the northern and eastern parts of the archipelago. It follows that snowfall will be heaviest along the western side of Baffin Bay and less by a factor of 0.6, along the northwestern section of the archipelago. Local wind conditions and regional snowfall distribution cause variations in the snow cover overlying the sea ice, with corresponding effects on sea-ice thickness.

Oceanographic conditions control the rate of ice growth through the availability of sensible heat, which limits ice growth. Any physical process that brings relatively warm, deeper water to the surface will thereby limit ice growth. Figure 13 shows the thermal structure of the water across Barrow Strait during the spring of 1984. Water, which resides near 100 m depth and is about 0.05 c° above its in-situ freezing point, extends to the surface along the south side of the strait, adjacent to Somerset Island. This persistent feature was recorded in previous years, when it was attributed to the upwelling of water flowing southward out of McDougal Sound (B. Bennett, personal communication).

Ice thickness is also plotted in the figure, and the ice overlying the warmer surface water is clearly about half as thick as the ice across the rest of the strait. No corresponding variation in snow-cover thickness or density occurred, and the thinner ice must be attributed to oceanic heat. If such zones of thin ice are consistently present, as appears the case, then knowledge of such regions is important in the selection of routes for ice-breaking vessels.

Very different conditions exist in the Beaufort Sea. When sea ice grows, the salts in seawater cannot be accommodated within the ice matrix and for the most part are rejected into the underlying water, thereby increasing the salinity of underlying water. This cold, dense water becomes unstable, and vertical convection occurs. If the water column is well mixed to the sea floor, then the water of excess salinity can reach the bottom of the sea and move laterally, as a density current, into deeper, subsurface water (Gade et al 1974). The continental shelf off the Tuktoyaktuk Peninsula in the Beaufort Sea is ideal for the production of such cold-density currents. The shelf is shallow, with the 50-m contour lying up to 125 km offshore. Furthermore, prevailing winds cause ice deformation, with ridge and lead production (Hibler 1979) forming new ice and brine.

The absence of shallow shelves throughout the Arctic Archipelago inhibits transport of cold, brine-enriched water below the surface layer. Melling et al (1984) describe the movement of warmer Atlantic water from the Arctic Ocean eastward through the deep western channels of the Queen Elizabeth Islands. Within the archipelago, heat from Atlantic water diffuses upward as the water moves slowly southeastward, progressively warming the halocline situated near 100-m depth. As the water moves into shallow passages between the islands, such as Penny Strait, Hell Gate, and Cardigan Strait, the flow accelerates and becomes increasingly energetic, mixing water from the warmed halocline into the surface layer. Topham et al (1983) found the water within one narrow channel to be as much as 0.2 c° above its in-situ freezing point. The availability of large amounts of sensible heat prevents or seriously retards the growth of sea ice throughout the arctic winter and results in a recurring area of open water called a polynya.

Polynyas: Polynyas are found throughout the Arctic Archipelago, as shown in Figure 14. Two mechanisms lead to their formation. Lewis (1979) describes "convective polynyas," where sensible heat associated with the upwelling of warmer water inhibits ice formation, and "latent heat polynyas," where no sensible heat is present in the surface water but new ice is removed mechanically from the area by strong currents and wind. In general, a given polynya will owe its existence to a com-

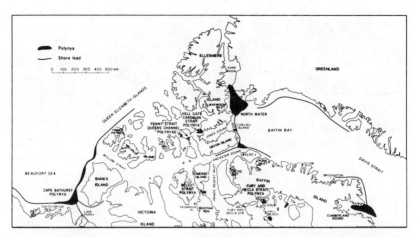

Figure 14. Recurring polynyas in the Beaufort Sea, Arctic Archipelago, Baffin Bay and Davis Strait. Adapted from Smith and Rigby, 1981.

bination of these two mechanisms, although one or other process may dominate.

Polynyas that vanish beneath a thin ice cover during winter, and areas where thin ice persists, will be the first areas to become free of ice during spring melt. Within the Arctic Archipelago, specific areas routinely become ice free weeks before the decay of ice in the region becomes well established. Examples are shown in Figures 15 and 16. Figure 16 illustrates the ice-free channels that develop early on between several islands off northwestern Bathurst Island. Cameron Island, the most northerly, is the site of potential oil production. An ice-breaking vessel en route to the loading site on Cameron Island during summer may occasionally encounter impassable ice at the north end of Byam Martin Channel, west of the islands shown. The regular early disappearance of ice between the islands, however, offers a good alternative, if not preferred route, to the Cameron Island site. Knowledge of areas of persistent thin ice is even more important in route selection for ice-breaking vessels navigating in arctic waters during winter, when thin ice cannot be visually detected and remote sensing, using infrared sensors, is effective only if the ice is a few millimetres thick.

Two very large polynyas, North Water and the Cape Bathurst Polynya, exist in northern Baffin Bay and the southeastern Beaufort Sea respectively (see Figure 14). North Water is thought to owe its existence to the prevailing wind, which breaks up and removes new ice from the area (Kupetskiy 1962). As ice is swept away, new ice is continuously and rapidly forming. Because thin ice grows more rapidly than thick ice, annual ice production in the polynyas greatly exceeds that in a static

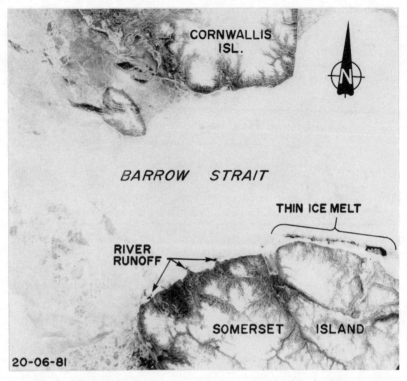

Figure 15. Satellite image of Barrow Strait showing early ice decay due to river runoff and thin sea ice.

ice sheet. Kupetskiy (1962) estimates that where 190-cm-thick first-year ice would normally form, continuous removal of ice results in production of 840 cm of ice. A good understanding of the physical processes controlling the extent of such large polynyas does not exist and must await extensive winter field investigations. Large polynyas are particularly important in the removal of sea ice in spring, because the open water and thin ice permit solar radiation, wind, and waves to initiate rapid ice disintegration.

Managing Sea-Ice Growth: It is possible to manage sea-ice growth by using engineering techniques to alter and control the distribution of available sensible heat in the underlying water column. This is practical only over relatively small areas but can limit ice growth near marine terminals. In order for a vessel to berth alongside a jetty, the ice must be less than a few tens of centimetres thick to allow the ship to manoeuvre. At least two types of systems have been proposed to inhibit

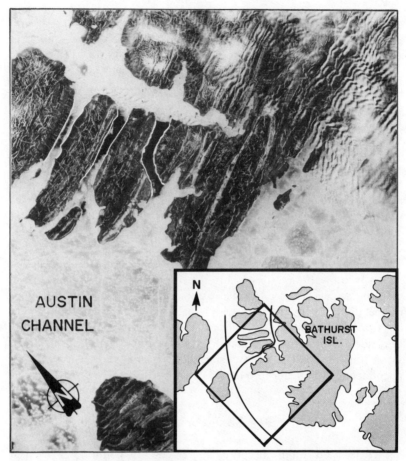

Figure 16. Polynya occurring in late spring in the channels to the northwest of Bathurst Island. The insert shows two alternate routes from Viscount Melville Sound to the Sverdrup Basin. The eastern route would take advantage of thin ice and early open water.

ice growth and could be used in combination (Haggkvist 1981). The "bubbler system," involves release of a continuous train of bubbles from a network of pipes, which often rest on the sea floor below the area to be controlled. The rising bubbles entrain sea water and lift the warmer bottom water to the surface, thereby maintaining a reservoir of sensible heat to inhibit ice growth. Such systems have been used effectively in Scandinavian countries (Carstens 1977). In the Canadian Arctic, substantial annual sea-ice growth renders bubbler systems ineffective.

An alternative ice-management scheme – using sea water as a coolant and thereby warming it – was proposed as part of the Arctic Pilot Project. It involved the loading terminal for liquefied natural gas (LNG) at Bridport Inlet, on Melville Island (D. Miller 1980). Attempts to remove ice there by ice-breaking tugs were considered counterproductive, as mechanical breaking and displacement of ice would increase ice production from a normal 2.2 m to about 6 m. Ice management is intended to reduce thickness but not free an area of ice completely. An open-water area would allow heat to be wasted through heat lost directly to the atmosphere and would result in unwanted ice fog.

In the scheme proposed for Bridport Inlet, sea water would be taken as a coolant into the LNG liquefaction plant, where it would be warmed from $-1.8°c$ (in winter) to $8°c$. The warmed sea water would be pumped to the wharf area and distributed below the ice through a series of nozzles, which could be used selectively to thin ice over an area of 200,000 sq m, or about ten times the area of the proposed LNG tanker. Calculations predict a 7-m-thick surface layer at $1.5°c$ (D. Miller 1980). The warm plume released into the ocean will be effective only if it is naturally buoyant, requiring that the discharged water be less dense than the ambient water. As the density of sea water alters little with temperature near its freezing point, small changes in water salinity will be critical in determining its density. To ensure that the warmed plume does not ineffectively sink away from the ice, the salinity of the intake water must be minimized, by using the less dense surface water or by introducing fresh water into the cooling system before release (Bennett 1978).

Sea-Ice Decay

Sea ice decays through the complex interaction of a number of physical processes, as shown schematically in Figure 17. The most important process is solar radiation, which increases rapidly with the lengthening of daylight during spring. At 75°N, for example, the sun is above the horizon for eight hours on 1 March, twelve hours on 21 March, and twenty-four hours by late April. The lengthening of daylight coincides with a high occurrence of cloud-free skies, resulting in high insolation at the sea surface. Air temperatures rise to above freezing, and ice growth ceases. As spring progresses, absorption of short-wave radiation by snow cover, sea ice, and underlying water increasingly promotes ice decay.

The ratio of reflected to incident solar radiation is called the albedo of the reflecting surface. A perfect reflector, for example, would have an albedo of 1. New fallen snow has an albedo near 0.9, and little

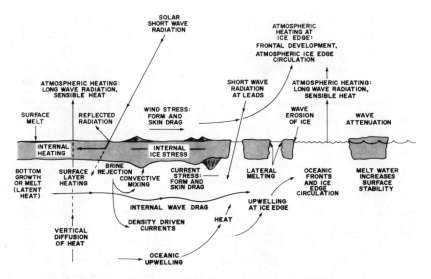

Figure 17. A schematic presentation of the physical processes involved in the atmosphere-ice-ocean system.

radiation will be absorbed by the snow and underlying ice. As air temperatures rise above freezing, the snow surface begins to melt and the albedo decreases. The albedo of snow can therefore vary rapidly in time in the order of 30 per cent, as snow changes between frozen and melting states. During the melt period, Langleben (1969) measured albedos that went from 0.5 to 0.2 as puddling on the ice surface increased. Surface melt-water effectively absorbs solar radiation, and surface ice-melt, once begun, can reach a few centimetres a day.

Incoming short-wave radiation during the polar summer is in the order of 300 watts/m². As surface albedo decreases, more radiation will penetrate the sea ice, where it is attenuated as it is transmitted to the sea beneath. The attenuation depends on the ice's optical properties, which are affected by air bubbles, ice-crystal size (Weller 1969), and brine pockets. Because of its higher salinity, first-year sea ice has more brine pockets and is more opaque than multi-year ice, and its attenuation coefficient is consequently higher.

Surface albedo can be drastically reduced artificially to promote early ice melt. For a few years Dome Petroleum Ltd placed a swath of coal dust on the snow surface leading from overwintering drillships in the Beaufort Sea to the edge of the landfast ice, so that it would move the vessels into open water at the earliest possible date, to commence off-shore drilling. This procedure quickly melted the snow cover and weak-

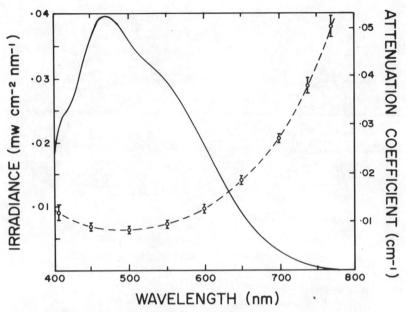

Figure 18. Transmitted spectral irradiance (solid line) beneath melting 1.2-m thick sea ice and the associated irradiance attentuation coefficient (dashed line) as a function of wavelength. After Roulet et al, 1974.

ened the underlying sea ice (J. Steen, personal communication). This procedure was stopped after arrival of the ice-breaking workboat *Canmar Kagoriak*.

The amount of solar radiation penetrating sea ice to be absorbed by the underlying sea will vary widely with thickness and physical, mainly optical, properties of the ice and overlying snow cover (see Figure 18).

Open water speeds the disintegration of sea ice and may exist during spring in the form of polynyas or leads. Leads in the pack ice develop when pack ice diverges. They are common in the Beaufort Sea, Baffin Bay, and, to a lesser extent, in the landfast ice of the Arctic Archipelago, where, for example, recurring leads stretch between headlands across Barrow Strait in May. Shore leads will develop in response to river runoff, as illustrated in Figure 15. Rivers carry relatively warm, silt-laden water that runs beneath or, in some cases, onto the sea ice. Runoff from land, being fresh and less dense, remains on the surface adjacent to the ice, where solar radiation is absorbed rapidly by the silty water, accelerating local ice melt.

The albedo of open leads is near 0.1, much smaller than that of melting ice. Leads therefore absorb solar radiation more efficiently and play a role in ice decay proportionally much greater than their areal extent.

Substantial energy is absorbed in the first few metres of the water column, and the warmed water melts adjacent ice through lateral meeting, and to a lesser extent bottom melting, thereby increasing the lead size.

As sea ice melt advances and runoff occurs, a surface layer of less saline water is developed. Horizontal density gradients become established at ice edges, resulting in the development of ocean frontal systems (Muench 1983). Wind-induced upwelling at the ice edge also produces oceanic fronts. Upwelling is caused by sudden wind stress on the sea surface at the edge of stationary or relatively slow-moving ice. Surface water is forced away from the ice edge to be replaced by warmer, more saline and nutrient-rich water from a greater depth. The enhanced supply of nutrients in the ice-edge zone can increase biological activity. Upwelling has been reported north of Spitzbergen by Buckley et al (1979) and in the Bering Sea by Alexander and Niebauer (1981).

The atmosphere also responds dynamically as air flows across an ice edge. Andreas, Tucker, and Ackley (1984) describe the heightening of the atmospheric boundary layer downwind from the ice edge. The air encounters greater surface roughness or drag coefficient among the ice floes. Air circulation changes as frigid air flows off a continuous ice sheet: water that is losing heat to the atmosphere forces it aloft, where it may develop cellular air circulation, similar to a sea-breeze (McPhee 1980). The resulting horizontal change in atmospheric pressure constitutes a frontal boundary. An off-ice breeze generated by the increased pressure gradient over the ice is turned to the right by the Coriolis force, resulting in an air-flow parallel to the ice edge.

Physical processes in the atmosphere and ocean in the marginal ice zone help explain the advance and retreat of polar ice edges. In the past few years this subject has received considerable scientific attention, through both numerical modelling and field measurements. Most of this activity has been brought together within the marginal ice zone experiment (MIZEX), which is focused primarily on the major global ice edges of the North Atlantic, Antarctic, and the Bering Sea. Investigation of marginal ice zones in both the Beaufort Sea and Baffin Bay may help determine the applicability of larger-scale ice-edge dynamics being investigated at the world's major ice edges.

Ice Break-up

As sea ice is weakened by spring warming and melt, atmospheric and oceanic forces fracture the ice sheet into smaller pieces more easily, and the ice becomes more mobile. The wind and currents apply stress against the upper and lower ice surfaces through surface drag, caused by friction and by form drag against ridges, ice keels, and so on. If the wind or currents vary horizontally, then a shearing force may cause leads to form and adjacent ice sheets to move relative to each other. In the water beneath the ice, lee waves may be formed on the pycnocline, as water flowing beneath an ice keel accelerates to pass beneath the obstacle. These lee waves travel on the pycnocline as internal waves and impart an additional drag against the ice. The freshened surface layer during spring partially decouples the surface layer from the water below. Floes consequently move more easily under the wind's influence, as less of the underlying water column is dragged along by the moving ice.

When the combined forces of the atmosphere and ocean are off the ice edge, floes are dispersed into the open water. Wind stress on the unconfined ice floe is transferred to the water beneath, thereby affecting the current speed and direction. In adjacent open water, wind affects the surface current directly.

The effect of vessel penetrations into the pack ice on ice-edge stability has been raised at environmental assessment reviews (FEARO 1980). It was suggested that vessels entering the ice near freeze-up or break-up would cause ice sheets to break off from landfast ice, thereby altering the position of the ice edge. It was argued that any resulting change of the ice-edge position would fall within the naturally occurring variation. While this may well be true, the redistribution of the ice edge resulting from vessel transects is not a random process and must always increase the areal range of open water. The true extent of such an impact, if any, is likely to be resolved only by monitoring several transects of the ice edge over time.

Modelling Ice Behaviour

Sea Ice: The drift and deformation of sea ice depend on the balance of forces acting on the ice. These forces include air and water stresses or drag, ocean currents, ice strength, and ice rheology. Any numerical model of ice behaviour consists of a complex set of equations, which include variations in both time and space. In order to consider air and water stress, the boundary layers at the ice surfaces must be described for a variety of conditions of surface roughness, including mixtures of ice floes and open water in varying proportions. Models normally use an idealized boundary layer, employing air and water drag coefficients determined from limited field experiments. Wind and current velocities are chosen in a manner appropriate to the scale of the process being considered. The most important components are the Coriolis force, air and water stresses, and ice interaction.

Internal ice stress and its relationship to ice deformation are an important but extremely complex parameter that must be included in a complete model of sea-ice dynamics. In dealing with the response of ice to internal stress, either a viscous or plastic rheology or a combination of these must be chosen, to give realistic behaviour of ice under shear and compression. The strength of sea ice and its variation with thickness and floe size are important.

Models dealing with ice dynamics in large areas – the Arctic Ocean or Beaufort Sea, for example – over several seasons must accommodate a variety of conditions. For example, ice strength changes with seasonal and diurnal temperature; ice thickness and the ratio of ice to open water vary seasonally; ice piles up in ridges and rubble mounds and interacts with shallow seas and shorelines; first-year ice and multi-year ice have quite different mechanical properties; frontal passages with high winds may cover a relatively small portion of the modelled area but must be spatially resolved in the model, as frontal circulation affects ice dynamics (Agerton and Kreider 1979).

Models of ice behaviour can be classified into two types: research and operational. Research models are designed to assess the sensitivity of ice behaviour to changes in various parameters, in order to explain phenomena such as climate (e.g. Semtner 1976), the effect of the ocean on sea-ice drift (e.g. McPhee 1979), and ocean circulation in the Arctic Basin (e.g. Hibler and Bryan 1984). Ice behaviour models can facilitate long-term predictions of ice conditions during an upcoming season of shipping and offshore activity. Prior knowledge of good and bad ice years would have tremendous economic benefits to off-shore industry, particularly in the Beaufort Sea, where ice conditions are notoriously variable from year to year. From 1955 to 1974 there were ten "good," four "fair," and six "poor" ice years for navigation (Milne n.d.). Long-range shipping forecasts will depend on development of accurate, long-term forecasts of weather and oceanic conditions.

Operational models are intended to predict ice or iceberg behaviour, usually drift trajectories, in real time, to assist marine activities such as offshore drilling. All models of ice behaviour contain assumptions and approximations, to allow for efficient and economical computing and insufficient data and understanding of physical processes.

A sophisticated operational model has been developed to support offshore drilling in the Beaufort Sea (Leavitt et al 1983). It provides daily forecasts of trajectories of significant ice features, such as large multi-year floes, ice velocities at selected points, wind fields, ice motion, and ice characteristics. It is intended to give advance warning of the encroachment of large multi-year floes, which can interfere with drilling operations. The model performs reasonably over periods of twenty-four hours or more. However, some short-term events are not well modelled, probably because of insufficient short-term data, the spatial resolution (grid size) of the model, and deficiencies in model physics.

Icebergs: Icebergs infest the waters of Baffin Bay, eastern Lancaster Sound, and the Labrador Sea. Icebergs, which can weigh up to 30 million tons, constitute a hazard to marine activities and to offshore hydrocarbon exploration in particular. In these areas operational models are required to provide real-time predictions of daily iceberg motions over a few tens of kilometres. Iceberg drift models incorporate forces caused by wind drag, water drag, and the Coriolis force (S.D. Smith and Banke 1981). Difficulties in modelling arise from the extensive sail and keel areas and the vertical extent and complexity of iceberg shapes.

Adequate forecasting of iceberg drift will require accurate forecasts of wind, currents, and waves. Meteorological forecasts are probably sufficiently accurate. Data are required on the variation of current with depth and over relatively small (10-km) spatial resolution. Wave data are required for wave amplitudes greater than about 3 m. Information is also required on iceberg size and shape. Present technology cannot generate such complete data for use in real time. Technological development in acoustic oceanography may overcome this limitation in the next decade.

Underwater sound is the only practical form of energy available for remote sensing beneath the ocean surface within useful ranges. Sensing may be either passive (extracting information from ambient noise) or active (through the transmission and receiving of acoustic signals). The latter method uses echo-sounders, correlation sonar, doppler, and other devices. Acoustic equipment located on the sea floor or on a vessel may, with sophisticated data processing, reveal fields of current motion, resolved both horizontally and vertically to the sea surface; sea ice and iceberg motion; wind speed and direction; and amounts of rainfall (D. Farmer, personal communication). A prototype correlation sonar for arctic use is being developed at the Institute of Ocean Sciences, in Sidney, British Columbia.

DATA AND ASSESSMENT

Data on the marine physical environment are used to assess both the impact of the environment on a given project and the impact of the project on the environment. The former type usually involves assessment for engineering purposes and is used to specify the design and operational criteria of vessels and other marine structures and, ultimately, the financial viability of the project. Involved are such parameters as the strength of ice features and extreme values of winds, waves, and ocean currents; risk analysis deals with the statistical chance of encountering such physical phenomena as extreme ice features, severe storms, or extreme waves.

The latter type of assessment – of a project's impact on the environment – has a quite different orientation. Examples of effects are possible retardation of sea-ice break-up because of the presence of several closely spaced artificial islands in the Mackenzie Delta, or the effect of repeated vessel transects on ice-edge stability. Projects may affect the physical environment in two ways. First, they can alter the environment so as to inhibit the project itself in some way. As examples, removing the sea-ice cover during winter will result in ice fog, thereby reducing visibility; repeated passages of an ice-breaking vessel may increase local ice production. Second, a project-induced change in the environment may have social and economic effects that can be positive, negative, or both. For example, a change of ice patterns could alter the migratory route of a marine mammal species, thereby increasing harvest in one arctic community at the expense of another. The latter type of environmental impact tends to be the greatest source of public concern.

The Data Base

The collection of data on the arctic marine environment is a difficult and expensive process and has a marked seasonal bias. Sea ice can prevent passage of research vessels and will often damage or carry off moored instruments. During the fall, sea ice is often too thin to support aircraft landings; in the winter, daylight is insufficient to permit off-strip landings. For these reasons oceanographic data in the Arctic Archipelago are more likely to be gathered in the spring. In the Beaufort Sea and Baffin Bay, where vessels can operate, more data are collected in the summer.

Within many regions of the Arctic, modern oceanographic instruments can only barely measure several oceanographic parameters, such as speed and direction of current and horizontal differences in sea-water density. For example, current speeds for significantly long periods can be below the threshold of speed-sensing rotors on recording current meters, usually near 2.5 cm/sec. Analysis of data from such a record will produce mean current speeds and directions that are skewed toward higher current speeds. The gradients of properties such as salinity and temperature determine the transport of properties such as heat and mass, yet in some parts of the Arctic, the Arctic Archipelago for example, very small gradients severely test the stability and precision of instruments. Frigid arctic temperatures also affect electronic data-recording equipment, and water samples, collected for calibration purposes, often

Table 2
Quality of Marine Data

Rating	Beaufort[a]	NW Pass[b]	Q. Eliz. I[c]	Baffin[d]	%
0	3	0	3	0	<1
1	19	6	21	3	8
2	36	62	42	105	42
3	55	62	54	74	42
4	21	3	7	4	6
Total number of data sets	134	133	127	186	

a Cornford et al 1982
b Birch et al 1983
c Fissel et al 1983
d Birch et al 1983

freeze, thereby invalidating oceanographic data. In environmental impact assessment these limitations need to be considered, and accuracies should be specified and substantiated. All too often this does not happen.

Not all data of a given type are compatible or of equal utility or quality. Data are collected by a variety of methods and for different purposes. Salinity data obtained in support of biological research may be accurate to 1 per cent, which is adequate for the original purpose, but are likely of little use to the physical oceanographer studying dynamic processes. Documentation of data with such information as location, time, accuracy, and resolution also varies widely yet is essential for assessment purposes. As the assessment process often makes use of all available data, the quality of all data must be taken into account.

The Institute of Ocean Sciences has recognized this need and is preparing and distributing a series of reports that compile and appraise physical, chemical, and biological marine data by region. These reports strive to identify all available data and to indicate the quality of each data set according to the following scheme: 0: data found to be wrong; 1: data suspect; 2: insufficient information; 3: data internally consistent; and 4: data consistent and standardized, and intercomparison with other data sets possible.

The quality of marine data for the Beaufort Sea, Northwest Passage, Queen Elizabeth Islands (i.e. northern Arctic Archipelago), and Baffin Bay, based on the above criteria, is shown in Table 2. These numbers indicate that about half of the available physical oceanographic data appraised are at least internally consistent, with no evidence of errors.

Oceanographic parameters are both spatially and temporally variable. Spatial scales vary from a kilometre or less to ocean-basin widths, and time scales from hours to inter-annually. Examples of variability are seasonal variability of ocean currents (Figure 3) and inter-annual variability in ice-edge position in the Beaufort Sea (Figure 9). Variability often makes accurate description or prediction of an oceanographic parameter difficult. Environmental impact assessment can be seriously faulted if it is based on data for which the degree of variability is not stated or is unknown.

An adequate description of site-specific physical marine data should include at least the average or most likely value, the extreme values, and the period or cycles of variation. For some parameters, such as tides, this is not too difficult, but for many others a comprehensive description is often not feasible, as the data required and the length of sampling time are often prohibitive. Regional descriptions, which involve increased spatial variability, compound the problem further. The time from initial project conception to environmental impact assessment is usually relatively short, so that the initiation of data collection during the first stages of a project may not produce a satisfactory data base, particularly for parameters having major inter-annual variations. There is clearly a need for long-term sampling of key parameters such as currents, ice regimes, and water levels, beyond project-specific requirements. The need for long-term research on the arctic marine environment has been recognized by the Environmental Assessment Review Process (EARP) panels for the Arctic Pilot Project (FEARO 1980) and Beaufort Sea Hydrocarbon Production and Transportation (FEARO 1984).

Future Research and Assessment

In general terms a number of common research needs are apparent. Year-round and longer-term data on non-tidal water levels, currents, water structure, and waves are required to provide better statistics on extreme values and the magnitude and frequency of variations. These data are also required to support research into the causes of long-term changes in the physical environment. More accurate meteorological data and long-term forecasting will be required to improve prediction of waves, storm surges, and operational ice-motion models. A greater understanding of atmosphere-ice-ocean interaction in the marginal ice zone will improve ice-motion models and prediction of the behaviour of ice edges and polynyas.

The process of assessing environmental impact requires improved methods to ensure consistent and accountable assessments, with clear limits of validity, for managers, regulators, the public, and politicians. Each assessment requires a clear audit trail of decision-making in each step of the process. It is important to know, for example, whether a particular assessment is based on an educated guess or on a widely accepted interpretation of a substantial data set. The limits of accuracy, resolution, and precision of data should be reflected in the assessment based on that data. When possible, a standard assessment method should be used by all participants in any assessment process, to ensure consistency.

Expected technological advances within the next decade should permit much more comprehensive measurement of data. Accordingly, these data will allow much improved assessment of environmental impact. We have seen the potential of acoustic technology to provide data on horizontally and vertically well-resolved fields of motion, sea-ice and iceberg motion, wind speed and direction, and so on. Hurlburt (1984) states that the application of technology should permit altimeters aboard satellites to measure changes in elevation in the order of 5 cm, which, together with scatterometers, will improve measurement of such parameters as sea-surface elevation, surface wind speed, and wave heights. Improved sea-surface temperature measurements from satellites can also be expected. The spatial resolution of satellite sensors to an order near 10 km will allow resolution of meandering currents and ice motions, which account for much of the variability in surface motions. Continued advances in computers' capability will lead to eddy-resolving basin- or global-scale models that can use the improved satellite data.

CHAPTER THREE

Marine Mammals and Ice-Breakers
Brian D. Smiley

INTRODUCTION

After this brief introductory look at marine mammals in the Arctic, arctic shipping, and popular attitudes to marine mammals, this chapter examines in detail contact between mammals and ships, demonstration projects and their role, research and its problems, and the future of both research and practice.

A chapter about marine mammals is essential in this book about ice-breaking transportation in Canada's Arctic. If there were no or few seals, whales, and polar bears in the Northwest Passage and Beaufort Sea, there would probably have been little environmental concern about Petro-Canada's ice-breaking Arctic Pilot Project (APP) or Dome Petroleum's oil-tanker proposal. Certainly, hundreds of thousands of seabirds which nest at several locations in the Passage can be billed as bigger stars on the arctic marine shipping stage. But issues about seabirds focus usually on unpredictable accidents – groundings, fires, or explosions – that cause oil releases to the sea. An oil-spill of catastrophic proportions is low in probability – one chance in thousands, or only a handful of times during the lifetime of a project.

In contrast, the routine and predictable operations of the ice-breakers pose threats to sea mammals, particularly to thousands of ringed seals, narwhal, and white whales, which inhabit the ice and waters in the proposed shipping corridors. A major accidental oil-spill will only exacerbate this situation, since ringed seal pups and polar bears insulated with white fur (like seabirds warmed with feathers) are sensitive to fouling, wetting, ingestion, and thermal stress.

Further, the hunt of marine mammals is central to the well-being of the Inuit, probably more so than their seabird harvest. Then, too, it is hard to imagine that the amount of attention captured by arctic shipping proposals would have been as great if one species, the bowhead whale, was not considered rare and endangered, numbering only several hundred or several thousands in the eastern and western Arctic, respectively.

In this chapter, I discuss some scientific perspectives about the assessment of ice-breaker impact on sea mammals. Most attention is given to the environmental impact reviews of the APP and the Beaufort Sea production and transportation proposal, highlighting the reviews' debates about whales and seals.

The environmental assessment and study of year-round ice-breaking traffic demand careful integration of the work of marine mammal biologists, hydrocarbon chemists, underwater acousticians, physical oceanographers, ice climatologists, naval architects, sea mammal hunters and many others. The viewpoint expressed here, however, is that of one person: an impact assessment practitioner and scientific reviewer who has written some, and read most, of the environmental impact statements and supportive literature concerning arctic shipping projects proposed or initiated over the past five to ten years. In 1979, he prepared an early environmental assessment in an "attempt to inform government scientists of various arctic disciplines about the nature of the [APP], and about its major environmental concerns and outstanding data gaps ... prior to the formal application of the APP to the National Energy Board and Federal Environmental Assessment Review Office" (Smiley and Milne 1979).

The reader will, no doubt, be tempted to join in this mental and emotional wrestling about mammals. A warning may be necessary – the struggle to understand with any degree of confidence will be frustrating, tiring, and unending. The following discussion cannot boast to be a thorough or even balanced presentation of viewpoints within the scientific field, industry, and public communities, nor are the perspectives developed necessarily endorsed by all or most biologists, engineers, and others. The subject matter is complex and technical, and difficult for anyone to describe plainly and for others to appreciate fully.

"There are more questions than answers," and "The more you find out, the less you know," in this interdisciplinary topic of science, technology, and assessment. Be prepared for tough questions, which are addressed with soft answers: "We don't really know ... " and "It depends ... "

Arctic Shipping

When approaching the issue of arctic shipping and sea mammals, we probably think that few Canadians have first-hand ship experience in the ice-covered waters of the Northwest Passage and Beaufort Sea. This perception is not true. Ship traffic has been routine in arctic Canada and has increased considerably in recent years. Pharand (1984), in his comprehensive overview of Northwest Passage navigation, cites forty known complete transits over an eighty-year period (1903–83). Vessels recorded include a fishing boat, a schooner, a tug, a Boston whaler, several yachts, a drill-ship, various research and survey ships, numerous ice-breakers, and some submarines. The flags of Canada, Japan, the Netherlands, Norway, and the United States waved over these men and women embarking on exploration, adventure, patrol and sover-eignty expeditions, hydrographic surveys, resupply assistance, scientific research, or mere short-cuts to or from home port.

From 1977 (when the NORDREG reporting system was established) to 1981, 129 ships made partial transits; 108 were Canadian, 21 foreign. Two of the Canadian ships were naval vessels, and the others included ice-breakers, survey vessels, fuel tankers, dry cargo ships, bulk-carriers, tugs, and drill-ships. The tankers took fuel to Rea Point on Melville Island for Panarctic, and the bulk carriers transported ore from the Nanisivik and Polaris mines. All the foreign ships, except three yachts, were tankers and bulk-carriers. Not included in this transit list are, for example, the suspected passages of nuclear submarines and the recent arctic shipping enterprise – the Bent Horn Oil Project, involving the first tanker export of crude oil from Cameron Island, through Barrow Strait and Lancaster Sound, then south down Baffin Bay.

Many hundred kilometres to the west, Dome, Gulf, and Esso have assembled a large fleet of vessels over the past decade, in support of offshore oil- and gas-drilling. For example, in 1985, there were at least sixty-six vessels, including six dredges, three drill-ships, four ice-break-ers, one tanker, two seismic boats, and fifty supply boats. Their traffic movements around the Mackenzie estuary and between offshore sites and shore bases were reported as totalling over 2,600 individual trips from June to December (Norton and McDonald 1986). In September 1985, the tanker *Gulf Beaufort* lifted anchor from the Amauligak oil discovery on the Mackenzie Shelf and departed westward to Japan, carrying the first-ever load (316,579 barrels of crude oil) exported from the western arctic offshore.

Clearly many different types of vessels have passed through arctic waters under varying circumstances and for many purposes, even though the numbers of people involved are few compared to other Canadian coastal waters. One must conclude that several generations of arctic marine mammals inhabiting the Northwest Passage and Beaufort Sea are already acquainted with large boats and ships, even during the icy seasons of fall and spring. If there have been any repercussions for sea mammals arising from this acquaintance, we do not know. Concerns about ship-mammal interactions did not arise until the late 1970s, and monitoring of ship passages under ice conditions is even more recent.

However, no one is experienced with year-round ice-breaking in the Canadian Arctic, as first proposed by Petro-Canada in 1978. Historically, for eight or nine winter months each year, arctic channels and straits have remained isolated from the noise, presence, and pollution of ships. The Beaufort Sea proposal forecasts, by the year 2000, between sixteen and twenty-six arctic class-10 tankers – each 0.8 km long, with 100,000-hp engines, carrying 200,000 tons of crude at 23 knots (open water) or 6 knots (3-m-thick ice). The project has naturally stimulated tremendous interest, emotion, and debate.

The first ship capable of year-round touring of the Canadian Arctic will probably be the Canadian Coast Guard's $0.5-billion arctic class-8 ice-breaker, to be built some day in Canada. "Class 8" means that the ship can operate all year in fifteen of sixteen Canadian marine zones and from 1 July to 15 October in the sixteenth zone, the Queen Elizabeth Islands, where Panarctic Oil has discovered oil and gas. Its capabilities are to maintain 3 knots in 2.5-m-thick pack ice and to ram multi-year ice up to about 8.7 m thick. The ice-breaker will carry 10,000 tons of fuel, enough for about three weeks (20 per cent reserve) at full power on the six diesel or gas-turbine engines, which generate 101,000 hp. With enough supplies for 175 people (including 60 scientific and military personnel) for nine months, it will have a theatre, saunas, a swimming pool, two helicopters, a hovercraft and landing barge, two all-terrain tracked vehicles, two 55,000-pound deck cranes, six laboratories, and a National Defence communications centre. The operating cost of Polar 8 is estimated at $250,000 a day during its varied year-round arctic missions (*Western Report* 1986).

Diverse Opinions on Marine Mammals

Almost everyone, it seems, whether a biologist or not, has an opinion about marine mammals and their protection and management. Large numbers of written interventions to project reviews from government, industry, and the public express concern about sea mammals. During the public hearings and review of the APP and Beaufort Sea proposals, about one-half to two-thirds of intervenors mentioned marine mammals as an issue. People seem compelled to express their observations, cite scientific evidence, and recommend more studies.

Maybe this is unavoidable: "Conflicting opinions, attempts to equate economic, social and moral values, misinformation and the impact of emotional arguments all make a consensus on seal and whale problems difficult to achieve. A divergency of philosophy plus the dearth of factual information concerning marine mammal biology often results in disagreements between the "experts" (Terhune 1985).

One seldom encounters such emotion and disagreement with other environmental issues. Because whales and seals are viewed as sentient and intelligent mammals, people feel strong empathy and sympathy toward these creatures.

A recent survey by Walter and Lien (1985) asked four thousand Canadian students in grades five and nine about the marine environment. Students generally felt very strongly about the ocean: 90 per cent indicated that the sea was "useful," 85 per cent "nice," and 83 per cent "important." Students had strong attitudes about marine wildlife (Figure 1). Their favourite sea animals were seals, dolphins, and whales, scoring high (over 80 per cent in the "love" and "like" categories). Salmon, trout, and sharks were also common favourites. The reasons given for choosing a favourite sea mammal were: it is "gentle and harmless" (physical, 29 per cent), has "spirit and special personality" (spiritual, 27 per cent), and it is "nice to have living" (aesthetic, 20 per cent).

Of course, such attitudes are predominantly those of southerners and do not necessarily represent those of children living along the arctic coast. The Inuit may have stronger likes and dislikes, together with the mixed utilitarian and cultural views of sea-mammal hunters.

MAMMALS AND SHIPS

Simply put, year-round vessel traffic in the Arctic will affect marine mammals in three ways: collisions, interference, and contamination. Whether or not these effects will be serious enough to impair the health and survival of the animals themselves or, in turn, hinder their successful harvest by Inuit offers some of the most frustrating questions for Canadians. For example, after examining the evidence and writing synoptic reports on the effects of vessel traffic on arctic marine mammals, two well-known Canadian biologists stated:

It is difficult to see how one could attribute changes in population numbers to the effects of vessel traffic with long-term monitoring of all the local ringed seal populations through the Northwest Passage. This would not be easy to accomplish considering the difficulties in obtaining food observational data, collecting adequate unbiased samples, and constructing meaningful population models. It is clear from studies on the ringed seals and the better known harp seal that unless one can sample a population adequately, it will be impossible to predict the outcome of a change in man-induced mortality, whether imposed by hunting or by some perceived environmental stress. (Mansfield 1983)

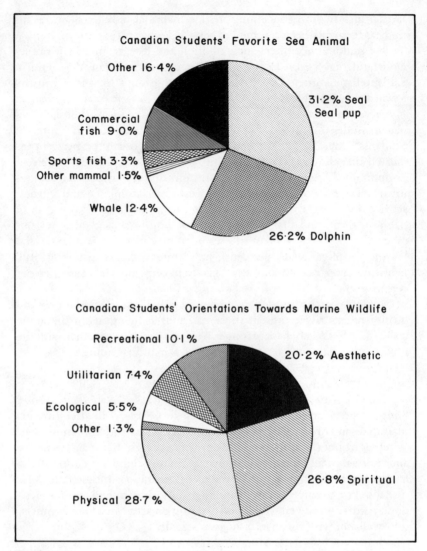

Canadian Students' Favorite Sea Animal

Other 16·4%

Commercial fish 9·0%

Sports fish 3·3%
Other mammal 1·5%

Whale 12·4%

31·2% Seal
Seal pup

26·2% Dolphin

Canadian Students' Orientations Towards Marine Wildlife

Recreational 10·1%

Utilitarian 7·4%

Ecological 5·5%

Other 1·3%

Physical 28·7%

20·2% Aesthetic

26·8% Spiritual

Furthermore:

It is generally conceded that it is virtually impossible to detect the effect of an oilspill in marine conditions unless it is of catastrophic proportions. This is particularly the case with respect to fish and marine mammal stocks where the effect on recruitment may only be apparent after a time-lag of several years. This being so for the highly studied east coast of Canada, it is even more difficult to expect valid statistical evaluation in the Arctic where, not only is the biology of the system completely unknown, but the great difficulties of enumerating stocks are frequently compounded by high natural fluctuations in abundance and migration pattern. The wide distribution and poorly known behavior of most marine mammals [make] it exceptionally difficult to quantify the stocks. Even more difficult is the problem of obtaining representative biological samples for the determination of age (compounded by the difficulties inherent in the aging procedures themselves) and reproductive cycle; for bowheads, samples are impossible to obtain in Canada as all hunting is banned. (L. Johnson 1983)

Before conceding that the scientific solution is "impossible," we should examine briefly some existing evidence and speculative theories about confrontations of ice-breakers and sea mammals.

Collisions

When whales or seals hear or see an approaching vessel, one would presume (and correctly so) that they dive or swim out of its path and avoid getting hit by the ship's hull and propeller. But large supertankers with high speeds in open water present a hazard, particularly for the slow-moving baleen whale. Patten, Samaras, and McIntyre (1980) reported, in a six-year study, eight deaths from fourteen collisions between ships and whales (mostly the slow-moving grey whale) in southern California waters. Weighing this evidence, Mansfield (1983) concluded: "It would be difficult to extrapolate from these figures, which represent an unknown proportion of the collisions of a fairly abundant species along a busy coast, to the situation in northern Canada where the population of the only large arctic whale, the bowhead, numbers several thousand in the Beaufort Sea and only a few hundred in Baffin Bay and Lancaster Sound; but occasional collisions might be expected."

A greater threat of collision is directed toward newborn ringed seal pups and their mothers, hidden in lairs excavated on snow-covered sea ice. During March and April, ice-breakers pressing through the fast ice of Barrow Strait, where birth lairs tend to be very common, could crush or wet the nursing pups (T.G. Smith and Stirling 1975). Until the pups develop an insulating layer of blubber, they rely on their dry white coats for warmth. The passing vessel may also cause flooding of the lairs along both sides of the ship's track, either by surging water up the ice breathing hole or by spreading water over the ice surface (LGL 1986). The APP (1981a) calculated that about 1 per cent of ringed seal pups born annually in Parry Channel (Northwest Passage) would be killed by eight passages of two ice-breaking tankers during this critical period. This is about the same number harvested each year by Inuit of nearby Resolute Bay. Obviously, heavier traffic would kill more seals, especially if the federal Department of Transport's predictions for shipping activities throughout arctic waters were ever to prove true (i.e., 74 grain ships, 69 ore carriers, 208 supply vessels, and 546 oil and gas tankers by 1995).

The seals' birth lairs are well hidden and camouflaged, fortunately from eyes of preying polar bears, but unfortunately from sight of the bridge of an ice-breaker. To locate and map birth lairs along the shipping corridor each spring would be an "impossible task." To date, the most effective technique is deployment of a black labrador bitch, who can smell and point out the under-snow lairs (T.G. Smith and Stirling 1975).

Interference

This is the most difficult section of this chapter to compose. The topic is complicated, we lack scientific and technical background knowledge, and all conclusions are speculative and open-ended. It is tempting to state merely that whales and seals may be disturbed and displaced by ship noise and physical presence, that their communications and echo-location can be hampered by ship noise (particularly low frequencies), and that their ice habitat will be harmed by ice-breaker passages.

These postulates are easy to state, but difficult to argue and extremely difficult to verify. Yet, second only to oil spills, noise-related effects on whales have dominated environmental impact assessments dealing with arctic shipping. This was the case during the Environmental Assessment and Review Process (EARP) review of the APP and Beaufort Sea oil production and before the hearings of the National Energy Board and the Senate's Special Committee on the Northern Pipeline. (As discussed later, the uncertainties and zeal surrounding ship/mammal noise helped the concept of phased development, using demonstration or pilot projects, gain wide and ready acceptance.)

In February 1981, Petro-Canada sponsored a workshop called "The Question of Sound from Icebreaker Operations," to which it invited over thirty physicists, biologists, and engineers. The company was responding to criticism by the EARP panel, the Danish Ministry of Environment, and others, regarding the assessment of the shipping component in the APP Environmental Impact Statement. A biologist at the workshop outlined the problem:

We are faced here with the need to predict whether or not the sounds produced by the proposed ships will be damaging to the lives and ecology of 32 taxa of northern marine mammals ... along the entire track of the ship. As background, we need to describe the animals that live there, the sounds they make, the uses to which they put sound, and whether or not we know of circumstances in this litany that raise concern. The pinniped and cetacean species (all of which are vocal, many highly so) found in the area of the proposed route include: walrus, harbour seals, ringed seal, ribbon seal, grey seal, bearded seal, harp seal, hooded seal, bowhead whale, black right whale, blue whale, fin whale, minke whale, beluga, narwhal, Atlantic white-sided dolphin, white-beaked dolphin, harbour porpoise, Risso's dolphin, the Atlantic long-beaked whale, Sowerby's beaked whale, common dolphin, blue-white dolphin and bottlenose dolphin, and more. Whether these animals and their normal populations can maintain their normal feeding, social structure and navigation in the face of the noise threat under discussion here is unclear to me. (Norris 1981)

Even after two days of wrangling, participants were able to agree only that:

The species thought to have the greatest potential for interaction with arctic icebreaking carriers through the Northwest Passage and south route are narwhal which are deepwater animals widely distributed in small groups in May and June throughout the pack ice of the northern 2/3 of Baffin Bay, bowhead whales considered an endangered species, because of the large gaps in knowledge of this species and because of an overlap in low frequency ship noise and use of low frequencies for bowhead whale communication, about 50,000 hooded seals which concentrate in an annually variable whelping patch in Davis Strait in March, 150,000 harp seals which are present in Lancaster Sound in July and August. The generally widespread distribution of ringed seals, polar bears and bearded seals suggests that interactions between these species and LNG carriers is inevitable but is not expected to have a serious effect on their behavior or distribution. For such coastal species as white whale and walrus, which have specific summer areas away from mid-channel shipping lanes, potential interactions are much less likely. However, several unknowns remain. (APP 1981b)

These predictions apparently were welcomed and considered a significant accomplishment by the workshop's organizers: "I was very apprehensive about data gaps and the lack of information to allow us to make any discrete decisions. There are admittedly very large data gaps. However, despite that, I think there are indications, some from anecdotal accounts and some from new research, that it is possible to make some estimates of what these effects will be" (Møller 1981).

Theory and data about noise-related effects are too complex and involved to discuss much further here. W.J. Richardson et al (1983), an invaluable 250-page document, relies on over 400 references to describe the acoustic (and non-acoustic) effects of the oil industry on twenty-five marine mammal species, but it concludes: "As a general comment, the reader should be aware that there are very few unequivocal data on the effects of non-consumptive human activities on marine mammals. Thus, few clear-cut conclusions can be reached, although some suggestive inference can be made."

Existing data are a perverse concoction of acoustic physics and mammal biology, with biochemistry and animal physiology thrown in. Even the experts readily admit their unacquaintance with these interrelated disciplines, many completely outside their experience and training. Assessment panels, hearing tribunals, and most intervenors have been at an even greater disadvantage. Often, it was a struggle for members merely to comprehend the common units of sound measurement, such as a source spectrum level of 125 dB//(1 uPa)^2/Hz at 1 m and a frequency of 1000 Hz, or minimal audible angles of 13.5° at 8 kHz for pure tones.

But, fortunately, in most helpings of solid but bland science, there are some sweet raisins of exciting observation. One such fillip is the findings of an ambitious three-year study (1982–4) conducted to monitor the behavioural and acoustical responses of narwhals and belugas to the approach and ice-breaking activities of the cargo vessel MV *Arctic* and the ice-breaker CCGS *John A. Macdonald* – one of the few field experiments designed to test the arctic noise/disturbance issue.

The experiment was made possible by the plans of Canarctic Shipping Ltd to lengthen the shipping season by breaking ice to the Nanisivik mine in Admiralty Inlet of north Baffin Island, near Lancaster Sound. The hunters of Arctic Bay were concerned that these earlier-than-usual entries (late June) into the inlet's ice edge and fast ice would scare away narwhals and interfere with the hunt. The biologists observed:

The results of the 1983 study suggest that the sensitivity of the whales to ship noise is even greater than was indicated by the 1982 study. In 1983, visible changes in the (whales') surface behavior and audible changes in their underwater vocal activity indicated that whales were aware of the ship's approach when it was in excess of 80 km from the ice edge. A strong avoidance reaction by belugas was observed in 1982 when the ship was 35–40 km away, and in 1983 when it was 43–49 km away. In contrast to the beluga's flee response, narwhals exhibited a freeze behavior. Both species vacated the disturbance area well in advance of the ship's arrival. In both years, narwhals were the last to leave the area. Belugas consistently demonstrated a greater degree of sensitivity to disturbance than did narwhals. (Finley et al 1984)

A layman might conclude that the once-remote populations of arctic whales and seals will probably suffer extreme annoyance and, given time, serious behavioural or physiological effects as a direct result of year-round shipping by ice-breaking vessels.

A good scientist usually refutes and denies expression of such thoughts

– i.e. that whales and humans, both being intelligent and warm-blooded, would sense and react alike. However, in this issue of vessel-related effects, some experts have invited such comparisons. For example, during the Petro-Canada workshop on noise, a Danish physicist concluded: "The effects on humans cannot be directly applied to marine mammals such as whales and seals. Recommended limits for underwater exposure of marine mammals to low frequency (sound) and infrasound are completely uncertain. There are several reasons for not using the same limits as for human beings. However, these limits are the only ones available. The reader is free to make his own comparison" (Møller 1981).

Another often-discussed topic, indirect icebreaker–sea mammal impact, should be mentioned here. Ice-breaker traffic could change the sea-ice cover, even modify the patterns of break-up, freeze-up, and lead formation. Since seals feed and breed on this ice, and whales feed and migrate along its edges and leads, the stage is set for indirect consequences – ships affect ice, which in turn affects mammals. Much speculation, some common sense, and scraps of data suggest that: repeated ice-breaking on forming ice in early winter could deter some seals from establishing territories in the areas traversed by ice-breakers; ice-breaking could delay consolidation of the ice sheet or advance break-up of fast-ice edges, thus affecting the feeding patterns of some whales and seals; and ice-breakers' refreezing or closing leads could entrap accidentally some migrating and feeding narwhals and white whales.

The ecological significance of such interferences, either additive or accumulative, is not known. For an overview of these relationships and other links among marine mammals, sea ice, ocean, and ships, see LGL (1986).

Contamination

A large oil-spill from an arctic tanker is unlikely but certainly possible under circumstances of collision, fire, or grounding. Some sources suggest that a maximum credible spill of 260,000 to 1.4 million barrels has a probability of 1 in 1,000. Regardless of predictions, during the lifetime of a major shipping project, such as that proposed by Dome Petroleum, several large spills are inevitable. Reliability, risk, and probabilities of spills have demanded almost as much time and energy in northern public hearings as the noise issue discussed above.

"The definitive answer to the question – Will oil affect arctic marine fish and marine mammals? – is an unqualified yes" (L. Johnson 1983). Most biologists and impact assessors agree that oil can directly affect seals, whales, polar bears, and walruses in any of several ways: mortality, plugged body orifices, irritated eyes, impaired movement, reduced insulation of fur, thermo-regulatory stress, and inhalation and ingestion leading to physiological stress. Add to these the indirect effects, such as displacement from critical habitats, reduced reproductive success, increased susceptibility to predators, and reduced food availability, and one sees why many pages of statements and hours of public hearings have been devoted to the topic.

There are few records of arctic sea mammal deaths directly attributable to accidentally spilled oil. However, experimental evidence does suggest that seals and polar bears, when involuntarily immersed in crude oil, suffer harm. For polar bears, this "harm" is death, after licking fouled fur and ingesting large quantities of the oil (Engelhardt 1981). For ringed seals, the harm appears transitory – eye and ear inflammations and kidney disorders; if, however, the animals are already stressed and weakened, either by other man-induced perturbations or by natural ecological conditions, temporary oiling worsens the situation and causes quick death (Geraci and Smith 1976).

For world-wide accounts of the effects of past petroleum spills on marine mammals, the reader is referred to Duval, Martin, and Fink (1981). Post-spill findings, laboratory studies, and other evidence demonstrating biological effects of oil on whales, seals, and other mammals were reviewed by Smiley (1982), Geraci and St Aubin (1980), L. Johnson (1983), and many others. Ringed seal pups and polar bears are usually considered to be most sensitive to oil, because of their easily soiled insulative coats and/or likelihood of grooming and ingestion. Next in line are bowhead whales, which feed at the surface: oil could foul their baleen.

Responses about long-term and indirect effects often contain the now-familiar phrases "There is no clear evidence or specific information ..." and "One cannot exclude the possibility ... " For oil-related effects, there is a long list of qualifiers, for example: "highly dependent on the species and their general health, age and sex of individuals, as well as circumstances surrounding the spill such as amount and type of oil spilled, time of year, success of cleanup operation, and the duration and spatial extent of contamination" (ESL 1982).

DEMONSTRATION PROJECTS

During the past decade of uncertainty surrounding ice-breakers and marine mammals, industry and government began to advocate phased development, or cautious incrementalism, for the Arctic. Such an approach relies on controlled experiments, known as demonstration or pilot projects. These limited-scale operations, carefully designed and cautiously implemented, are expected to determine the best or preferred technology or procedures for mitigating adverse effects, while minimizing serious risks to life and limb and avoiding irreversible, widespread threats to the environment. "Mega-project" has become a nasty word and is no longer used to describe a development strategy for frontier areas of uncertain risk.

In June 1982, John Munro, federal minister of Indian affairs and northern development, told the Canadian Club: "We've decided to accelerate government research, planning and monitoring for northern hydrocarbon development in a comprehensive and controlled manner. The object is to put ourselves in a state of preparedness to allow initial production to begin from proven commercial reserves through demonstration projects. The controlled approach through demonstration projects will facilitate the introduction of special measures that may be needed to mitigate northern impacts and, at the same time, maximize northern benefits" (J. Munro 1982).

Another department senior manager told the Beaufort (Sea) EARP (BEARP) Panel: "The federal government position on Beaufort hydrocarbon development indicates government support for phased development – that is, carefully planned and managed development of proven commercial reserves initially through demonstration projects, as against massive and rapid development" (Faulkner 1983).

Roméo LeBlanc, federal minister of fisheries and oceans, wrote to John Roberts, federal minister of environment, in March 1981: "I am pleased that the proponent (Arctic Pilot Project) had the foresight to propose, and the (Beaufort Sea) Panel to support, a long term research program. More knowledge of the Arctic environment is needed before we can predict with any degree of confidence, the real impact of proposed energy development, as well as providing the base for the federal planning, regulatory and decision-making functions. This pilot project would provide a valuable opportunity to assess the environment effects of year-round Arctic shipping, and to devise suitable protective measures in anticipation of future expansion of Arctic marine activities" (LeBlanc 1981).

Fisheries and Oceans supported this approach for the Beaufort Sea: "Expansion of offshore production and transportation should be considered only after the demonstration projects have been shown to be safe, reliable and environmentally acceptable. We advocate a phased approach to year-round tanker traffic, (offshore production facilities and subsea pipelines), essentially involving a progression of demonstration projects, whereby key environmental and safety issues would be resolved in discrete steps, under conditions of controlled risk" (DFO 1984).

And industry has been quick to remind government and others of their endorsement of and commitment to offshore development: "Many intervenors (at Beaufort Panel hearings) have expressed a preference for phased development. The Federal Government has also supported this concept. Industry has indicated that phased or balanced development is a likely scenario. Phased development will build on the experience of previous activities. For example, our exploration programs have utilized experience to construct islands in progressively deeper water, learning more about interactions between man-made structures and the environment. This approach in a frontier area is also compatible with the socio-economic, environmental, technical and financial aspects of Beaufort development" (Dome, Gulf, Esso 1983).

Phased development was a product of uncertainty and inexperience about ice-breakers and marine mammals (and, of course, concern for crew and vessel safety). Petro-Canada first advanced the strategy – as exemplified by its staged proposal (the APP) to use ice-breakers to ship liquefied natural gas from Melville Island, through much of the Northwest Passage, and south to market. More recently, BEARP recommended that the government of Canada approve the use of oil tankers to transport Beaufort oil only if two stages were followed: first, comprehensive research and preparation; second, a two-tanker stage, to demonstrate environmental and socioeconomic effects within acceptable limits (FEARO 1984).

Taken at face value, this endorsement and consensus appear a rallying commitment to small-scale trials (with, one hopes, only small attendant errors) to gain on-the-spot experience. Yet misunderstanding, fear, and mistrust continue about any course of development that unfolds in phases and as demonstrations.

Differing Viewpoints

The following sub-section presents five different viewpoints, from industry, the Inuit, an interest group, an academic, and a scientist. It reveals that not everyone is comfortable with the pilot-project compromise.

The industrial viewpoint was summarized by D.M. Wolcott, chair of the executive committee, APP, in a presentation to the EARP technical hearings in Resolute Bay, April 1980:

The [APP] project permits incremental development. We think that the sensible approach to year-round marine transportation in the Arctic is to proceed one step at a time – from small scale pilot to a full scale system, if the pilot proves successful and if a larger system is warranted. This approach does not preclude a pipeline or full scale marine delivery system in the future.

The project, after initial study and consultation, does not seem to disturb, beyond acceptable levels, the relationship which exists between the Inuit and their environment or between the wildlife and their surroundings. The effectiveness of this project will enable rational judgements to be made concerning the development of other gas fields further removed and the development of other resources from the High Arctic. The project has been called a "pilot" because it is designed at the minimum scale necessary to prove the technical and economic feasibility of delivering Arctic Islands natural gas by ship. It is also termed a pilot because of its size compared to any full scale alternative for the delivery of Arctic gas, be it by pipeline or a large volume marine delivery system. It would be one-tenth the size of any other energy-related project being planned in the Arctic. This small scale would minimize construction delays, environmental impact and all the potential problems that have beset previous large megaprojects located in remote northern areas.

It is considered a pilot also because it may be discontinued and removed with minimal disturbance, should it be deemed necessary. Finally there are elements of uncertainty that must be evaluated. Knowledge gaps are present because we have only now completed the preliminary engineering phase of the project. We have used available literature, consulted with local people and in many cases conducted field studies to try to ascertain what impact the project would have on the environment, and vice versa. But in some very key areas, we will only know the impact after the pilot has been implemented. Because in real terms, we are exploring; no one has tried to build a project like this before. (Wolcott 1980)

A technical expert concluded: "This project is an innovative pioneering pilot project. It recognizes the unknowns associated with such a project in terms of year-round shipping in arctic waters. There is much to be learned, both now and during the operations of those ships. Only with the LNG carriers in the water can we better understand the effects of underwater noise from these ships" (Holman 1981).

In May 1982 three northern native groups – the Inuit Tapirisat of Canada (ITC), the Baffin Region Inuit Association (BRIA), and the Labrador Inuit Association (LIA) – submitted a brief to the Senate's Special Committee on the Northern Pipeline. The brief was probably one of the most powerful and well-articulated statements ever made about the suspicion and frustration surrounding the APP. The following excerpt illustrates the worries of the native people:

The APP means different things to different people. It is variously an energy project, a project to test novel technology, an economic opportunity for Quebec or Nova Scotia or, for the Inuit a crisis to be met. Is it really an experiment? Or will its approval lead to dozens of icebreaking tankers in Arctic waters? If it appears inevitable, should Inuit (notwithstanding their grave apprehension as to the environmental and therefore economic impacts of it) seek to participate in it? Or should they throw their all into stopping it?

The Government of Canada, when it decides at its various levels whether or not to give APP a green light, must understand what it is doing. The ITC, BRIA and LIA, and the Inuit who have given testimony in opposition to the APP, contend that the APP, of itself, poses unacceptable environmental risks, that there is a high probability that it will lead to other tanker projects, and that taken together, these projects will lead to environmental disaster ...

The Inuit of Canada wish to tell you what the APP means to them, the environment and to the Inuit way of life. The APP would represent the first commercial attempt to establish the capability of year-round shipping through the Northwest Passage, Davis Strait and the north Labrador Sea. Even with the technology of icebreaking tankers as yet untested, there are already more ambitious proposals for marine transport of hydrocarbons working their way through the system. It is obvious to everyone that the APP is only the thin edge of the wedge, and that while being packaged as a "pilot" project, there is no actual prohibition against larger, more ambitious marine transportation proposals until the impacts of the "pilot" have been assessed.

We do have the assurances from the proponents of APP that they will not affect whelping seals and that they will depend on officers of the Department of Fisheries and Oceans to help monitor the situation and to identify significant patches of whelping seals. Nevertheless, the Department does not have those criteria established and, when weighed in light of the ship operator's mandate, there is no doubt about how seals will fare against the always pressing priorities of time and money. (ITC, BRIA, and ILA 1982)

One well-organized and dedicated interest group, the Canadian Arctic Resources Committee (CARC), made an intervention in March 1980, part of which is as follows:

PetroCanada has stressed repeatedly the "pilot" nature of its proposal to move gas from the Arctic. It has claimed, quite correctly, that the APP would test the technical and economic feasibility of moving arctic resources by the marine mode on a year-round basis. The APP however is not a "pilot" as the word is commonly understood. Costing 2 billion dollars and involving 20 year contractual commitments, the project will not be changed or abandoned easily if significant technical, economic or environmental problems arise. If approved, the APP will represent a permanent commitment in its own right. The only questions its sponsors will debate will be whether to increment it at a later day by adding more ships and facilities.

The APP's true pilot nature lies in the development that it may precipitate or bring forward in time. As the first year round commercial marine venture in the Arctic, the APP will facilitate others, in the same way any pioneering venture tends to do. The assessment, approval and terms and conditions for this first arctic marine transportation system will thus establish the framework for all that follow. A successful Arctic Pilot Project will encourage other shipping ventures in the North. These ventures will undoubtedly attempt to replicate the successful aspects of the APP.

Basic to success is routing. If the APP's proposed routing is proved successful, others will quite rightly expect to follow similar paths and be subject to similar terms and conditions. Shipping corridors will thus become established. It would not be unreasonable to anticipate about 1000 transits a year through the North-west Passage by the year 2000, although the actual origins and seasonal preferences may change this order of magnitude. This is what PetroCanada's "pilot" project, if successful, will bring in its wake. (CARC 1980)

On matters of science and policy about arctic shipping, the opinions of university professors and others have been advanced and even solicited to provide yet another perspective, presumably independent of government, industry, and native communities. For example, after considering the costs and benefits of shipping through the Northwest Passage, Boothroyd and Rees (1981) concluded:

The question is not only whether to proceed with Arctic shipping, the question is also – at what speed? There are many advantages to a slow pace to industrial development and shipping. A slow pace would allow research to be taken on the many still unanswered questions about the environmental impacts of development and shipping, and on the mitigation possibilities before commitments are made to courses of action that could have major implications. A slow pace would mean that the inevitable mistakes that will be made are more likely to be small mistakes in terms of their environmental, social, economic, technological and intergovernmental implications. A slow pace would allow us to develop the navigation, surveillance, enforcement and cleanup systems necessary to minimize the risk of environmental damage occurring from industrial development and from shipping.

Scientists, too, are reluctant to place hopes on phased development:

The Beaufort Sea Panel reports their serious examination of the effects of icebreaking and echos concerns of DFO and other intervenors regarding the lack of understanding of the stability of fast ice regimes in deep water (Amundsen Gulf and Lancaster Sound, in particular). Recommendation #31 advocates research into the effects of icebreaking on ice regimes, #32 advocates research into the effects on snowmobile transits over sea ice, and #36 advocates research into the effect on seal behavior and mortality. The Panel proposes in Recommendation #22 that routine use of tankers only be permitted following a two-stage evaluation. I regard this as a very sensible suggestion, although I am skeptical whether the second stage, a two-tanker demonstration project, will provide answers to many of the topical questions of environmental impact – because of the difficulty of carrying out controlled experiments. Thus the requirement that "... environmental effects [be] within acceptable limits" before routine use of tankers rings somewhat hollow. (Melling 1984)

RESEARCH

Money and Experts

The most recent Directory of Marine and Freshwater Scientists in Canada (DFO 1986) indicates that about 2,200 professional people are engaged in research, development, consulting, management, and scientific services across the nation. However, closer examination reveals that Canadian scientists who are well qualified, properly trained, and available for arctic marine research probably number in the tens. To address even the major marine mammal issues over the next decade or two will take a large number of biologists, physicists, and engineers. This requirement has already been noted by some scientists:

Many of the recommendations and comments within the [Panel] report express the need for research, and the Department of Fisheries and Oceans is assigned its fair share of projects: oilspill trajectory models (Recommendation #3), oilspill cleanup, behavior, detection and effects (#8); tanker feasibility (#22); fate of marine pollutants (#29), effects of artificial islands on ice (#30); effects of icebreaking on ice regimes (#31) and on seals (#36); research on beluga whales (#38), narwhal (#39), sea mammal distribution (#45), behavioral response of sea mammals to vessels (#46), vocal communication of sea mammals and masking by ship noise (#47) and physiological response to ship noise (#48). The specific research recommendations are buttressed by Recommendation 60, which proposes a very major (15 year) Arctic research program by government.

I have no argument with the topics suggested for research ... However, it is obvious that the amount of research work is well beyond the capability of existing DFO Arctic groups, and perhaps beyond the capability of existing Arctic expertise in Canada, given the short (4 year) time frame suggested for many projects. (Melling 1984)

Even assuming arctic expertise capable of the research task, is there sufficient money? A group of about eighty US and Canadian experts on all aspects of noise, the arctic environment, and the oil industry convened in 1980 to assess the potential hazards of man-made noise associated with Alaska arctic wildlife and to prepare a research plan to secure the knowledge necessary to permit proper assessment of noise impact. Their rough estimate of the cost of such a plan was thirty million dollars over five years, assuming major co-operative efforts by government, academe, private, and industrial concerns (Acoustical Society of America 1980).

Similarly, after an in-depth review of the Beaufort Sea Panel report, DFO's Arctic Research Directors' Committee concluded that scientific investigation of the panel's recommendations dealing only with marine mammals, noise, and oil would cost thirty million dollars over five to ten years and require 250 person-years (i.e. fifty scientists working for five years). At present, even with accelerated baseline research under way for the next few years, the department's programs amount to about one-quarter of the effort needed (DFO 1985).

If Canadians are expecting scientific solutions to the issues of arctic shipping and marine mammals, time must be allowed for building and training the teams of experts, and millions of dollars made available as dependable funding.

Quality of Data

It is reasonable for everyone to expect that any information about sea mammals be complete and reliable – that is, consisting of data that are well maintained, well documented, accurate, and precise. Given the high cost in time, money, and expertise to obtain whale field measurements and observations, one might assume that sufficient care and attention would be devoted to quality control and documentation. Unfortunately, this has not been the case for the bulk of historical data on whales gathered in the Beaufort Sea region.

Norton, Smiley, and de March (in press) compiled and appraised all known (143) historical data sets about white whales, bowhead whales, and other cetaceans in the Canadian part of the Beaufort Sea, from whaling ship logs of the 1850s to present-day sightings from oil companies' drill vessels and aircraft. They developed a simple rating system for scrutinizing the methods-and-materials documentation, in order to judge objectively the reliability of the hundreds of measurements reported in each data set. The chosen rating criteria were commonsensical, elementary, and phrased as questions – for example: were the type, speed, and altitude of the aircraft recorded, or were all units of measurement defined such that another observer would be able to repeat the procedure and get the same value? If the investigators did not provide sufficient documentation to answer any or all such questions, measurements or observations were assigned a "2."

Of the nearly 1,700 different measurements available from all sources (published and unpublished, analysed or stored, public or proprietary, government- or industry-generated), 70 per cent were assigned this "2" rating, indicating insufficiency. This creates a dilemma for biologists such as impact assessors who must rely on historical data (often collected for different purposes) for predicting change and judging its significance. Without the instructions – details on collection, storage, analysis, and so on – it is difficult or impossible to interpret the pieces. Considerable argument usually erupts during any discussion of ship and whale interactions.

Quantity and Timeliness

Many people have grown accustomed to getting answers quickly, especially if there is enough money and energy devoted to a problem. This is a false expectation in scientific investigations of marine effects such as ice-breaking, noise, and oil-spills. The only norms of the arctic environment are extremes and variability. This not only makes interpretation of data very difficult, but also makes collection risky, frustrating, and often unproductive. Aircraft or shore-watch observations of whales are thwarted by fog, snow, waves, and darkness. Severe winds and cold restrict acoustic listening to sea mammals through the ice. Hours or days are spent waiting on the ice or waiting for the weather to change or equipment to be repaired.

The Arctic seems to resist those who come expecting to obtain data and leave in a week or two, or even a year or two:

Generally, very little is known about marine mammals. In attempting to assess the influence of noise, I suspect that we are attempting to describe the abnormal while the normal is still unknown. Examples in the open literature often suggest that noise will be detrimental to marine mammals and fish. The addition of noise to any system is virtually, by definition, detrimental. The presence of anecdotal accounts and conflicting information leaves discussion of the effect of noise open to argument. The eventual detrimental effects will possibly range from no observable (or actual) effect to extirpation of a species from a particular region. For the case of marine mammals, assessing the influence of various noises is or will be extremely difficult and costly. The study conducted by Reeves, Ljungblad and Clarke on bowheads in the fall of 1982 is an example of this difficulty. Over a period of 38 days, whales were observed for only

18 hr. 46 min. from a fixed-wing aircraft. Their findings were equivocal. (Brown 1980)

Or consider the Beaufort Sea Project. Unfortunately (but probably not too unexpectedly), the sea ice never moved more than 20 km off-shore during the summer of 1974, restricting two research ships to operating around Herschel Island. Most of the planned oceanographic stations could not be occupied. All this at a charter cost of $1.5 million.

Good data are hard to get and sometimes unreliable. This is a reality, and must be taken into account in any future demonstration projects. But how?

Ideal information is often lacking in practice, and we are faced with basing decisions on small samples of data which must be judged and interpreted largely on their own internal evidence. Many people feel that small samples are so inherently unreliable that it is best not to draw conclusions from them at all. Whether we like it or not, much scientifically relevant evidence that is available from the work of ourselves and others is, in fact, rather scanty. It is a counsel of despair to ignore it. And in any case, people will inevitably draw conclusions from rather inadequate material. So the most sensible procedure is to try to use methods of analysis that will tell us objectively what conclusions, however vague, can justifiably be drawn from the data. (Bailey 1959)

One such "sensible procedure" is the t-test, for common significance. Marmorek (1984) used it in constructing a simple mathematical model to show the relationship among sample size, observational and natural error, and the detectability of various-magnitude effects. He presented it at a workshop on systematic techniques for analysing uncertainty in determining ecological change. Although the analysis is neither precise nor entirely quantitative, it shows some serious shortcomings of a demonstration-project solution.

Marmorek's model assumes that, before introduction of an environmental stress (such as ten year-round passages of two 100,000-hp class-8 ice-breakers), the environmental variable being considered (e.g. number of seal birth lairs per 100 sq km) varies randomly around some mean value – that is, values for consecutive years are not correlated. Following application of the stress, the variable continues, it is assumed, to vary around some (possibly different) mean. The scientific challenge then becomes: for various combinations of sample effort, can a change in the mean be detected at a certain level of significance?

Some of the model's results, solved by exploring standard t-test tables and equations of variance with a microcomputer, are presented in Figure 2. The y (horizontal) axis of the diagram represents the sampling or

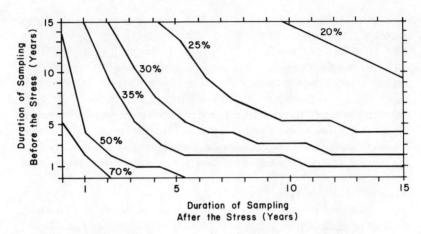

Duration of Sampling
After the Stress (Years)

observation effort before the stress, and the x (vertical) axis, the effort of sampling after the stress. The six contour lines represent different levels of change of mean (a so-called impact) that can be detected with particular combinations of sampling effort. (For this example, the coefficient of variation was set at 30 per cent and the cut-off for significance at 95 per cent probability).

Some interesting, even startling relationships can be seen. First, a large (70 per cent) reduction in the mean (probably catastrophic in real life!) could be statistically detected two years after the stress, even with no years of pre-impact sampling. Second, a moderate (25 per cent) reduction in the mean can be detected five years after the stress begins only if about fourteen years of sampling were carried out before the stress. Third, a small (less than 20 per cent) reduction in the mean could be detected, but only with decades of study before and after the stress is introduced.

Two general trends are also evident from this model: the effects of stress can be detected in the shortest time by maximizing the number of sample years before its introduction, and if there is relatively little sampling effort before the stress, it becomes more and more difficult to increase the sensitivity of detecting a change by extending sampling after the stress.

So, what do these models and statistics mean in the context of arctic demonstration projects involving marine mammals? It appears critical that, at the outset, everyone in industry, government, and the scientific and public sectors agree on the amount of environmental change (impact) which, if detected, will trigger a management decision about the project. For example, is a reduction of 10, 25, or 50 per cent in the sampled variable (such as number of narwhals in their summering fiord, or number of seal birth lairs per 100 sq km of fast ice) considered significant – or serious enough to the animals' population or the people who hunt them – to warrant rerouting ships, revising their timing, frequency, and speed, modifying their engineering, or even stopping their operation altogether? Without a prior agreed-upon level of significance to trigger such tough decisions, the necessary monitoring program cannot be designed.

Further, industry argues that, as a feasibility test of most technical and economic factors, a pilot project need operate only for three or four years before escalation to the next larger phase. As illustration, Panarctic's Bent Horn Oil Project made application in 1984 to quadruple its phase-1 shipment, of 16,800 cu m, after three years of operation, but now it plans to double shipment after only two years (IERC 1987). However, the shorter the period of demonstration, the longer the delay imposed on the project's start-up time. For example, if there already exists a three-year biological data set on seals or narwhals, an additional five years of pre-project sampling are still required (a total of eight years of before-stress sampling) if a four-year pilot project is to have any success in statistically detecting, with 95 per cent confidence, a 30 per cent or greater impact caused by shipping activities.

THE FUTURE

Patience and Trust

Offshore industry is probably correct in saying that only with ice-breaking tankers in the water can we better understand the effects of underwater noise and other disturbances from ships. That being the case, trusting agreements and patient co-operation are the challenge for all concerned parties. If arctic shipping is to commence year-round in Lancaster Sound, Beaufort Sea, and other ice-covered waters, some new ways of thinking and acting will be necessary to achieve the scientific objectives of a careful and cautious demonstration project for gaining knowledge and experience under circumstances of low risk.

There has been little appreciation of the Inuit hunters and trappers who, during their daily activities on the ice and ocean, make similar measurements and observations concerning ocean currents, ice climatology, and animal behaviour. The southern scientific community must begin to meld its data sets with those existing in the minds and memories of men and women in coastal arctic communities.

Using a humorous and all-too-true analogy, one observer likened the activities of researchers from government laboratories and from industry to those of other arctic wildlife:

Two waves of migration are observed by arctic people during the spring. First come the birds. Then come the researchers. They are a noticeable and significant social phenomenon as they load their equipment and gear into four wheel drive trucks, transfer it to float and ski planes, stow it on boats and head to their field camps, research institutes and study sites. Their presence across the Yukon and NWT is monitored with interest and can generate strong feelings and strongly ambivalent attitudes. A fear of some residents is that they will be left on the outside, looking over the fence ... as ships move through the Arctic Archipelago on a year-round basis. [They are] ... particularly sensitive to the possibility that the Canadian Arctic can be seen as a vast laboratory filled with social, biological and physical science data to be exploited at will by colonially-minded, southern-based experts who have very little real investment in the social and political aspirations of northerners. (McTiernan 1983)

In the future, biologists, oceanographers, and Inuit hunters must be prepared to collect, analyse, interpret, and apply data together. Trust and patience are central to these relationships. Otherwise a pilot project will do little to demonstrate safety and protection to those "left looking over the fence."

Commitment and Funding

The federal Department of Fisheries and Oceans (DFO) advocates that "Beaufort Sea hydrocarbon development should occur at a rate commensurate with the demonstration of safety, reliability and environmental acceptability based on available scientific data and understanding" (DFO 1985). To tackle just the shipping component and effects on marine mammals, the estimated requirement is two dozen competent scientists working with complete dedication for ten years, with an annual budget of three million dollars (DFO 1985). This effort should produce some results. But will they be pertinent, and who will use them? How, when, and where will these results be incorporated into decisions?

The Department of Indian Affairs and Northern Development (DIAND) is following a policy of cautious phased development, as discussed earlier. "An essential component of such an approach," the department stated to the BEARP Panel, "must be an environmental research and monitoring program, fully integrated with the exploration and development scenario, and designed to provide an early warning of any unexpected environmental effects" (DIAND 1983a). To this end, the Beaufort (Sea) Environmental Monitoring Program (BEMP) was initiated in April 1983, to determine the most significant environmental component to be affected by the industry, to recommend and continually evaluate mitigative measures, and to adapt future research and monitoring projects to reflect information gained and changes in the industrial strategy.

Phase I of this program was carried out under contract by well-known environmental consultants. Two interdisciplinary workshops and a series of technical meetings were attended by experts and resource managers from government, industry, universities, and the consulting community. The objectives of these activities were to distinguish between perceived and real effects, based on present knowledge of industry proposals and ecological processes. The goal was to represent the majority viewpoint of a range of disciplinary specialists with experience in the Beaufort Sea region. A method called "Adaptive Environmental Assessment and Management" was selected to draw together the parties involved. Simulation modelling of the Beaufort ecosystem and hydrocarbon development scenarios was attempted, primarily to force each specialist to view his/her individual area of interest in the context of the whole system (BEMP 1985).

Fifteen "Valued Ecosystem Components" (VECs) were selected – resources or environmental features that are important locally, nationally, or internationally and in management or regulatory policy – bowhead whales, king eiders, polar bears, and arctic ciscos among them. Probably the most challenging task was to develop and evaluate critically hypotheses about effects relying on known or suspected links between the industry and VECs. Some of the twenty or so hypotheses propounded were simple, but most were more complicated food-web interactions. Nine hypotheses and about fifty linkages dealt directly with marine mammals. For those hypotheses considered valid by the cross-discipline workshop teams, monitoring plans were proposed. In some cases, such as the bowhead whale studies, DIAND undertook to see that the monitoring and basic research proceed and the results be made available in following years, for further validation of impact hypotheses.

An ongoing research-planning forum like BEMP must accompany any future demonstration projects. Workshops must become a dynamic teaching and learning ground for all parties, ensuring that decisions can be acted on by everyone. More active participation by senior managers of government, industry, and Inuit will be demanded; they must get together with researchers for several days each year, if new scientific findings are to help direct the scale and scope of phased shipping activities. To date, BEMP has cost $60–80,000 each year, excluding conduct of its recommended studies. Many days of thought, travel, and review by some of Canada's most respected arctic scientists and engineers, regulatory managers, and northern inhabitants have been devoted to this exercise. But even greater involvement of Inuit representatives is warranted.

The Environmental Advisory Committee on Arctic Marine Transportation sponsored a small-scale version of BEMP, to look at monitoring

of present-day shipping activities through the Northwest Passage. The
remarks of industry representatives suggest the value of such meetings:
"It is a useful early warning mechanism in terms of proposed studies
and/or developments; it is effective in bringing a large number of players
together who otherwise would not undertake such dialogue; and it has
a great potential for focussing study needs with a view to producing
results for decision making" (Pessah and Robertson 1986).

However, future BEMP-like planning and assessment programs may
cost $100–150,000 annually during a major ice-breaking pilot project.
Such money would be well spent if results included useful, timely, and
reliable data for co-operative sea mammal research and monitoring.

Education

Much of the confusion and argument about effects on sea mammals
stems from poor understanding of the ocean, its basic life forms, and
its biophysical processes. Nevertheless, large segments of the general
public and, indeed, the scientific community consider that they have
valid philosophical viewpoints about tankers, oil, and whales. This leads
to more problems in assessing impact and in making policy. Conser-
vationists support the best use of the resource; preservationists tend
toward non-use. But accompanying rationales are often based on misconceptions about the ocean's "ecology."

The problem begins in school. In addition to the four thousand Canadian students surveyed in 1985 (mentioned above) two hundred teachers of grades five and nine were also asked for their opinions about
marine education. Almost all teachers (89 per cent) perceived a need
for more such education in schools, and about 70 per cent of the students
considered it important to study environmental issues in school. In fact,
among students choosing a career related to the environment, 40 per
cent elected to work in an ocean environment. However, today, only
57 per cent of schools teach students about the marine or aquatic environment. About 70 per cent of the students surveyed could not understand a basic marine food chain, did not realize that energy for marine
life comes from the sun, and did not know what plankton was. Only
16 per cent of schools use materials designed for the specific curriculum
area, and "the effectiveness of those [materials] that do exist may be
questioned as they do not appear to relate to any increase in students'
knowledge about the sea. It is evident that future Canadians will know
little about the ocean" (Walter and Lien 1985).

Similarly, Canadian students of bygone days, many of them now
attending public hearings, preparing interventions, and reading impact
statements, possess both this enthusiasm and this incomprehension.

From stewards of the sciences, additional resolve and exercise are demanded:

For both physical and biological applied problems in the marine environment, long-term progress is clearly going to depend on the proper division of effort between learning how the ocean works and making the measurements necessary to meet, as well as possible, immediate needs. With finite resources, we need in each situation to carefully consider the most effective balance between using available data, collecting basic datasets, carrying out process oriented experiments, and developing theoretical and numerical models. Perhaps the most effective way to progress is to insure the development of expertise in industry, government and universities so that those involved can rely on expertise and good sense where all else fails. (Needler 1980)

In the future, such expertise also must be fostered among northerners themselves, thereby adding to their present wealth of good sense. This goal has stimulated discussion about establishment of a northern research institute:

Scientific knowledge provided through data gathering, analysis, interpretation and synthesis into environmental impact assessments is the principal means used by government, industry and northerners in directing the course of resource development. At some point, judgements or decisions are made on the basis of this information. The effectiveness of these decisions depends on the certainty and quality of the information provided. [In this regard], the concept of a northern research institute deserves special attention as a focussing mechanism for providing a stronger basis for arctic decisions and for considering the northerners' aspirations. A scientific centre would foster the growth of a resident scientific/technical community and better informed people, by giving ready access to information, training in research, assessing impacts of projects, providing fora for seminars, lectures and meetings, sponsoring students for further education, and even orienting southerners to northern ways. (Pallister 1983)

Pallister concludes: "The University of the North concept once discussed is perhaps on too large a scale at this time to be established. Rather, a pilot or demonstration project merits serious consideration." What a noteworthy idea – to demonstrate the value of a northern research team in monitoring a year-round arctic shipping project meant to demonstrate safety and protective measures!

CHAPTER FOUR

Seabirds and Environmental Assessment
Richard G.B. Brown

The Canadian Department of the Environment has been in existence since 1971. One of its many functions has been to assess potential effects of proposed industrial development on the Canadian environment. In 1973 Environment Canada set up its Environmental Assessment Review Process (EARP), under the Federal Environmental Assessment Review Office (FEARO), to ensure that environmental impact is taken into account in the planning and implementation of federal projects. There are comparable environmental review processes for projects that fall under provincial or territorial jurisdiction.

Twenty-five EARP panel reports were published up to July 1984 (FEARO 1984). Two deal with hydroelectric projects, two with seaport or airport construction, four with nuclear power or uranium refining, six with railroad or highway construction, and eleven with discovery, production, and/or transport of oil and natural gas. Impact assessment of proposed hydrocarbon development projects is even more extensive than these numbers suggest: at earlier stages, initial screening and, if appropriate, initial environmental evaluation are required for every off-shore drill-site, and even for areas to be covered by seismic surveys. This practice reflects concern for the effects of spilled oil on beaches, fishing gear, and seabirds and other wildlife.

The effectiveness of EARP may be examined with particular reference to potential hazards that the offshore oil industry poses to seabirds and other waterfowl. My perspective is that of a research scientist working for the Canadian Wildlife Service (CWS), Environment Canada, on the seabirds of Atlantic Canada and the Arctic. My experience with EARP has gone from collecting baseline data on seabird populations and distributions, through reviewing consultants' reports and advising industry, to attending public presentations to EARP panels, both as a CWS "expert witness" and as a panel member. I have also had experience with oil-spills, both as a member of Regional Environmental Emergency Teams (REETs) and in estimating seabird mortality in the field. Since FEARO and the higher levels of EARP are discussed elsewhere in this volume, I shall deal mainly with preliminary aspects of the process.

SEABIRDS, OIL, AND ENVIRONMENT: PROBLEMS

We humans are curiously attracted to birds. The persistence of bird-hunting and the popularity of bird-watching in urban Western societies are both indications of this. The attraction is so hard to explain rationally to non-enthusiasts that it must reflect some very deep psychological need. Economics has little to do with it. The value of the ducks that a hunter brings home is far outweighed by the money spent, and discomfort endured, in pursuit of them; we bird-watchers do not even have a bird in the pot at the end of the day. Certainly the dollars-and-cents value, as meat, of ten thousand dead ducks and seabirds washed up after a major oil-spill is small compared to the costs to the fishing and tourist industries of fouled gear and beaches. Even so, birds are the most conspicuous victims of oil-spills, and their intangible, perceived value is very high indeed. The oiled bird on the shore has become a powerful symbol of the concern we feel over the increasing destruction of our natural environment, by oil pollution and by other means. The birds killed by oil from the supertanker *Torrey Canyon* in 1967 (Bourne, Parrack, and Potts 1967), and from the blow-out at a drill rig off Santa Barbara, California, in 1969 (Straughan 1971), helped set off the wave of "environmental consciousness" of the late 1960s.

Seabirds are important in EARP also because they are not just potential victims but may be environmental monitors. Seabirds are unusual among marine organisms in that they can be studied fairly simply, without expensive apparatus. Moreover, they may be used to monitor much more than the mere occurrence of oil-spills. As animals at or near the top of the food web, their fluctuations in numbers can be used as an index of the "health" of organisms at lower trophic levels, even of the marine ecosystem as a whole. It was, for example, reproductive failures in such birds as brown pelicans (*Pelecanus occidentalis*), ospreys (*Pandion haliaetus*), and common terns (*Sterna hirundo*) that first drew our attention to the toxic side-effects of the chlorinated hydrocarbons and heavy-metal compounds used in agricultural pesticides (e.g. Risebrough 1969; Fimreite et al 1971; Fimreite 1974; Peakall 1975; Bourne 1976; Ohlendorf, Risebrough, and Vermeer 1978). Vermeer (1976) and Vermeer and Peakall (1977), among others, have stressed the value of population censuses of waterfowl and seabirds in monitoring chronic pollution by oil and other chemicals in the Canadian marine environment. Gas chro-

matographic "fingerprinting" of oil taken from seabirds has been used
to identify the source of the oil (e.g. Brown and Johnson 1980; Levy
1980).

Oil in the sea affects a bird in two ways (Bourne 1976; Ohlendorf,
Risebrough, and Vermeer 1978; Brown 1982). The most obvious is
through contamination of its feathers. A bird's body is insulated by
pockets of air trapped in the layer of down-feathers next to its skin.
The down is protected on the outside by contour feathers, each of which
consists of a set of barbs branching out on either side of a central shaft
(e.g. Rawles 1960). Each barb carries a series of hooked barbules that
bind it to the barbs on either side. This interlocking structure makes
the feather water-repellent and, combined with the overlapping contour
feathers, gives the bird a relatively water-proof outer layer of body
plumage. However, the fine structure of barbs and barbules is easily
disturbed and has to be maintained by preening. Oil breaks up the fine
structure and allows water to penetrate into the down and eliminate
the insulating air pockets. The water is cooler than the bird's body –
markedly so in Canadian marine environments – and the resulting rapid
loss of body heat must be countered by metabolizing the body's energy
reserves (Hartung 1967; McEwan and Koelink 1973; Croxall 1976). Up
to a point, the bird can preen off the oil, restore the fine feather structure,
and survive minor contamination (e.g. Birkhead, Lloyd, and Corkhill
1973; Bourne 1974). But a badly oiled seabird in a cold climate is unable
to do this quickly enough to halt the drain on its reserves. It is caught
in a vicious spiral. It can replenish its reserves only by increasing its
food intake – yet foraging efficiency is likely to be decreased by, for
example, the extra weight of water it must carry around in its feathers
(e.g. Holmes and Cronshaw 1977). When the fat reserves are exhausted,
muscular tissue begins to break down, and the process of physical
degradation becomes irreversible.

There is another way in which spilled oil affects birds: as they attempt
to preen their feathers, they inevitably swallow a certain amount of oil
(e.g. Hartung 1965). The toxic effects of this are not well understood,
but they are probably sub-lethal rather than lethal inhibition of the
excretion of sodium ions, for example (e.g. Miller, Peakall, and Kinter
1978; Peakall et al 1980), and of oviposition and egg hatchability (Ainley
et al 1981). Experiments have shown that ingestion of oil by young
seabirds depresses their growth rates (Miller, Peakall, and Kinter 1978)
– which might cause them to leave the nest very late in the season, or
underweight, in both cases reducing their chances of survival. Even
small quantities of oil, transferred to the egg experimentally or from
the parents' feathers, may be enough to kill the embryo (Hartung 1965;
Birkhead, Lloyd, and Corkhill 1973; Albers 1977; Szaro and Albers

1977; Coon, Albers, and Szaro 1979; White, King, and Coon 1979). However, the effects of oil ingestion are minor compared to those of oil on the body surface.

Major and minor spills of oil at sea date back at least to the late nineteenth century. Spillage of fuel oil and refined petroleum from ships sunk in the Second World War was undoubtedly very large, but its effects on the marine environment have not been reported in any detail. However, the frequency and severity of spills have increased sharply in the last twenty years, with the construction of large supertankers (e.g. Bourne, Parrack, and Potts 1967; Bourne 1976; Powers and Rumage 1978; Stowe and Underwood 1984).

The first Canadian experience of a major spill came when the tanker *Arrow* ran aground and sank in Chedabucto Bay, Nova Scotia, in February 1970 (Anonymous 1970–3; Brown et al 1973). Some ten thousand tonnes of bunker c fuel oil were released; about a fifth of this remained in the bay, doing considerable damage to the shoreline ecosystem and to fishing gear. The remainder drifted offshore in three large slicks and would have gone undetected if two of these slicks had not come ashore on Sable Island, the only land between Nova Scotia and Europe (Brown et al 1973). A minimum of twenty-four hundred birds – mainly common eiders (*Somateria mollissima*) and oldsquaw (*Clangula hyemalis*) – were killed by oil in the bay itself, and a minimum of forty-eight hundred – mainly murres (*Uria* spp.) and dovekies (*Alle alle*) – came ashore on Sable Island. Meanwhile a minor, little-publicized leak of no more than seventy tonnes of bunker c, spilled from the oil barge *Irving Whale*, killed at least five thousand birds in southeastern Newfoundland (Brown et al 1973).

Since then we have had two major incidents affecting seabirds and waterfowl. The tanker *Golden Robin* ran aground in the Bay of Chaleur, New Brunswick, in September 1974, and spilled about four hundred tonnes of bunker c; several thousand ducks, loons, and grebes were killed (Hildebrand 1980; A.D. Smith, cws, personal communication). In March 1979 the tanker *Kurdistan* broke in two in Cabot Strait and spilled some eight thousand tonnes of bunker c, fouling beaches and killing about ten thousand birds, mainly ducks and auks (Brown and Johnson 1980). Meanwhile, the chronic spillage of small quantities of oil continues (e.g. Tuck 1961; Austin-Smith 1968; Brown 1973; Hildebrand 1980). At the time of the *Kurdistan* incident, for example, dovekies went ashore on Sable Island, covered with fuel oil chemically distinct from that released from the *Kurdistan* (Brown and Johnson 1980; E.M. Levy, Bedford Institute of Oceanography, personal communication).

Such assessments of bird mortality are inevitably underestimated.

Oiled birds can be counted only when they go ashore; one counts the number of carcasses along a beach of known length and extrapolates this to cover the whole length of oiled shoreline (e.g. Brown et al 1973; Brown and Johnson 1980). However, dead birds sink to the bottom after they have been drifting for a couple of weeks, and perhaps only 20 per cent of birds oiled at sea ever reach the beaches (Tanis and Morzer Bruyns 1968; Hope-Jones et al 1970). Dead birds may be removed from the beaches by scavenging predators or by clean-up crews. Moribund birds often crawl away from the shore and into cover and so are not included in the counts.

It is difficult to assess the significance of oil-induced mortality to bird populations as a whole. Brown et al (1973) suggested that the *Irving Whale* incident may have killed 1 per cent of the common eider population wintering off Newfoundland. However, ducks have a rapid reproductive turnover, with a high adult mortality rate balanced by a relatively high annual production of young (e.g. Ashmole 1971; Perrins and Birkhead 1983). The reproductive turnover among eiders is low compared to that of other ducks, but even so they are potentially able to recoup their losses fairly rapidly. This is not true of auks, whose reproductive strategy is geared to a very low adult mortality rate under natural conditions, balanced by small clutch-sizes (usually only one egg per year) and a long adolescence before the birds begin to breed. This strategy fails in the face of additional, "artificial" mortality caused by excessive hunting, drowning in fishing nets, or oil contamination. The minimum number of about forty-eight hundred murres, dovekies, and other auks killed in the *Arrow* incident (probably 20 per cent of the true figure) is small when set against the annual Newfoundland hunters' kill of over 350,000 murres (Wendt and Cooch 1984) and the approximately 65,000 auks drowned off Newfoundland in gill-nets in 1980 (Piatt, Nettleship, and Threlfall 1984). But the cumulative effects of these "artificial" mortalities will inevitably reduce murre populations, not only of Newfoundland but also to those from Greenland and the Canadian Arctic that winter off eastern Canada (Gaston 1980).

Several features of the Canadian marine environment exacerbate the effects of oil-spills, both on the marine ecosystem in general and on seabirds in particular. The cold water temperatures not only accelerate loss of body heat in oiled birds; they also ensure that the oil remains liquid and therefore contaminating for a long time. The lighter fractions quickly evaporate in warmer seas, leaving a residue of relatively uncontaminating tar-balls (Levy and Walton 1976; see also Bourne 1976). Canada's climate, and the inaccessibility of much of its coastline, make cleaning-up operations very difficult. Oil can be frozen into pack ice and released long after a pollution incident, and both pack-ice and icebergs will also increase the chances of damage to, and spillage from, offshore oil installations.

Canadian geography adds another dimension to the problem. Oiled birds that go ashore along well-populated coasts, as in California and the English Channel, can be collected and cleaned of oil, though the process is slow and often unsuccessful (e.g. Clark 1978). There are few such places in Canada. Our coastline is so long and so sparsely inhabited that it is often hard enough to locate the spill itself, let alone the birds affected. Small spills are apt to go undetected, while the inaccessibility of the coastline often makes it difficult to estimate reliably the numbers of birds killed in major events. This is not in itself an insuperable handicap; in New Zealand collection of seabirds washed ashore after storms and oil-spills is something of a national ornithological hobby (e.g. Veitch 1978). Unfortunately this action is not emulated in Canada. Routine beach-monitoring by cws and by other ornithologists is on nothing like the scale of New Zealand, let alone northern Europe (cf. Bourne 1976, tables I–IV).

CANADIAN SEABIRD INVESTIGATIONS

If we are to assess the actual or potential effects of oil, or of any other pollutant, on Canadian seabirds, we must establish some baselines for the sizes and distributions of their populations and also for other key factors.

We cannot restrict these investigations to species or populations that breed in Canada, since non-breeding migrants occur in Canadian waters at all times of year. At least six million individual seabirds breed in eastern Canada, from southern Nova Scotia to northern Baffin Bay, and two to three million non-breeding sub-adults are also present off-shore (Brown et al 1975; Brown and Nettleship 1984a, b). But these are augmented in summer by the world population of at least five million greater shearwaters (*Puffinus gravis*) from colonies in the South Atlantic (Cramp and Simmons 1977). Roughly four million thick-billed murres winter off Newfoundland, approximately 27 per cent of them from colonies in western Greenland and the remainder from the Canadian Arctic (Gaston 1980). Banding returns show that the large population of dovekies that breeds in northwestern Greenland (at least fourteen million birds: Salomonsen 1979; Renaud, McLaren, and Johnson 1982) winters off Newfoundland and that we also receive Atlantic puffins (*Fratercula arctica*) from Iceland and Scotland, northern fulmars (*Fulmarus glacialis*) from Britain, and black-legged kittiwakes (*Rissa tridactyla*) from as far away as Spitzbergen and northwestern Russia (Tuck 1971). The implication of these complex and wide-ranging migrations is that pollution of Canada's marine environment may affect populations of seabirds that breed far from its shores.

Research on seabirds in eastern Canada has been under way since the 1920s, and the Canadian Wildlife Service (CWS), the federal agency responsible under the Migratory Birds Convention Act for conservation and management of waterfowl and most seabirds, has always played an important role. Investigations began with censuses and banding studies of breeding seabirds, mainly in the Gulf of St Lawrence (e.g. H.F. Lewis 1924, 1931; R.A. Johnson 1940; Fisher and Vevers 1943–4); with surveys of the birds' distributions at sea (e.g. Wynne-Edwards 1935; Rankin and Duffey 1948); and with the pioneer work of the late Leslie Tuck at murre (*Uria* spp.) colonies in Newfoundland, Hudson Strait, and Lancaster Sound (Tuck and Squires 1955; Tuck and Lemieux 1959; Tuck 1961). Until recently, however, knowledge about Canadian seabirds was far behind that available for the seabird populations of New England and parts of western Europe (e.g. Fisher 1952; Fisher and Lockley 1954; Drury 1973; Cramp, Bourne, and Saunders 1974). Canada is a large, sparsely populated country, much of it inaccessible until comparatively recently. For example, the important seabird colony on Prince Leopold Island, Lancaster Sound, was discovered in 1819 but was not visited by an ornithologist until 1958 (Gaston and Nettleship 1981). Other arctic colonies were not discovered until the 1970s (Nettleship 1974, 1980; Nettleship and Smith 1975).

The situation began to improve in the late 1960s, when CWS initiated a series of comprehensive surveys of the seabirds of eastern Canada. CWS surveys continue, augmented since the mid-1970s by the work of academic researchers and by studies carried out by consulting agencies for the oil industry. These investigations include locating and counting seabird colonies (e.g. Lock and Ross 1973; Nettleship 1974, 1980), quantitative shipboard and aerial surveys of seabird distributions at sea (e.g. Brown et al 1975; Nettleship and Gaston 1978; McLaren 1982; Orr and Ward 1982), and investigations of the migrations, feeding habits, and reproductive and survival rates of seabirds – issues of fundamental importance to an understanding of seabird biology in both "applied" and "pure" research (e.g. Gaston 1980; Montevecchi and Porter 1980; Gaston and Nettleship 1981; Bradstreet 1982a, b; Brown and Nettleship 1984a, b).

These investigations have made Canada a world leader in seabird research. These surveys were already under way before the boom in Canadian offshore oil exploration began in the mid-1970s and provided fairly realistic, if preliminary, data for EARP, right from the start. In the early 1970s, on the west coast, virtually nothing was known of the seabird populations when Alaskan oil-fields first went into production, and the development of the Prudhoe Bay field and its ancillary pipeline and transport systems proceeded without the information needed for a proper environmental impact assessment. The subsequent, very detailed surveys of Alaskan and Californian seabirds (e.g. Sowls et al 1978, 1980; Gould, Forsell, and Lensink 1982) are only now beginning to fill this gap.

EARP

If initial evaluation of a proposal by Environment Canada or other federal departments suggests significant potential environmental risks, the full Environmental Assessment Review Process (EARP) is begun (FEARO 1979b; Munn 1979; Waldichuk 1982). The proposers present a review, based on the literature, of what is known about as many aspects as possible of the environment at the site in question. This preliminary environmental impact statement is drawn up according to guidelines set by the EARP panel, a group of reviewers assembled by FEARO from government and the private sector. Until recently the gaps in knowledge, especially on the biological side, have been rather wide, and such a review has usually led to a series of surveys to remedy this. Once this has been done, a definitive environmental impact statement (or set of statements) is drawn up, summarizing the state of knowledge and listing potential hazards, along with a contingency plan which outlines emergency measures to be taken in the event of an environmental accident.

These documents are reviewed by the relevant environmental agencies, including CWS. The proposal then proceeds to hearings held before the EARP Panel, at which the agencies and the public are represented. The panel in turn reports to the federal minister of the environment and, where necessary, to the ministers of other federal and provincial departments (e.g. FEARO 1980, 1984).

For a projected offshore oil operation initial stages of the process – statement and plan – should consider several questions.

1. In what way is the environment likely to be affected? In offshore hydrocarbon exploration, polluting effects occur usually through discharge of crude oil from the drill-hole itself and of fuel oil, and possibly also drilling mud, from the drill-rig and its tenders.

2. Under what circumstances will these substances be released? Accidents and malfunctions in the drilling operation may lead to blow-outs of crude oil; bad weather, and collisions between various permutations of rigs, tenders, and icebergs, may lead to release of fuel oil.

3. What, in actuarial terms, are the chances of such accidental discharges?

4. What will happen to released material? Will it evaporate, sink, or drift? In all three cases, is it likely to cause immediate economic concern

to industry; for example, by fouling fishing gear or tourist beaches or by destroying fish, gamebirds, and other organisms important to commerce or leisure? What will be its effects, in the long or short run, on marine biological communities – on phytoplankton, zooplankton, ichthyoplankton, benthic organisms, seabirds, marine mammals, and so on.

For seabirds we need to know whether there is any threat to large, local concentrations of birds, at their colonies and elsewhere. More specifically, is there a threat to the diving species (auks, ducks, loons, and grebes) which are especially vulnerable to oil pollution? Such assessments must take into account seasonal changes in the birds' distributions. Size of oil-spill does not necessarily indicate the mortality that may result.

5. In the event of a spill, how quickly, and by what measures, may it be contained? For example, what is the contingency plan for cutting off the flow of a blow-out? What should be done with the spilled oil? Is it possible to destroy slicks by spraying them with detergents, or clean up contaminated shores or oiled seabirds? If possible, is it desirable? Would the detergent "cure" be worse than the oil "disease" (e.g. Smiley 1982; Wardley-Smith 1983)?

CWS has maintained its role in seabird research in Canada but now emphasizes long-term studies of seabird biology (e.g. Gaston and Nettleship 1981). Collection of most data on seabird distributions needed for EARP has been done by consulting companies under contract to the oil industry. Here, in theory, CWS's role is restricted to critical assessment of the consultants' work. This arrangement is the result of the federal government's "Make or Buy" strategy, and it has obvious advantages for the public purse.

However, it soon became obvious that this official "hands-off" policy could not be made to work in the small world of Canadian environmental ornithology. I could hardly confine myself to my theoretical role of assessor, when I myself had assembled much of the information that the consultants needed for planning their surveys (Brown et al 1975; Brown 1985). It was a waste of everyone's time to separate too rigidly the roles of the consultants and of CWS and could have led to uncomfortably adversarial relations. We therefore evolved a system of unofficial co-operation at all stages up to submission of the environmental impact statement (EIS). At various times in a project I would find myself advising both the consultants and the industry that employed them on seabird surveys in a given area, briefing the consultants on background information that CWS had on file, and commenting on the survey results as they became available. Only at the very end would I get down to my official task of reviewing the definitive EIS. I was never the Lord-High-Executioner; the choice of a consultant for that job was and must always be a matter for industry alone. But there were moments when I felt like the Lord-High-Everything-Else.

The restrictions imposed by the small size of the relevant scientific community also affect higher levels of EARP. For example, the Arctic Pilot Project (Northern Component) proposes the transport of liquid natural gas by icebreaker-tanker, from the High Arctic islands through ornithologically sensitive areas in Lancaster Sound and Baffin Bay (FEARO 1980). At the public hearings in 1980, I was disconcerted to find myself not only a supposedly neutral EARP Panel member, but the only available seabird "expert" as well.

These official and unofficial arrangements have worked quite well in practice. Some excellent studies have come out of many of the surveys carried out by consultants as part of EARP. On the ornithological side alone, one may cite the symposium issue of the journal *Arctic* on Baffin Bay and Lancaster Sound (e.g. Bradstreet 1982a, b; Bradstreet and Cross 1982; McLaren 1982; McLaren and Renaud 1982; Renaud and McLaren 1982; Renaud, McLaren, and Johnson 1982) and other studies by Bradstreet (1980), Orr and Ward (1982), and Orr et al (1982). Knowledge of the pelagic distribution and ecology of seabirds in eastern Canadian waters has advanced, especially through the use of aerial surveys, far faster than seemed possible only a decade ago. Compare, for example, the skimpy, ship-based coverage of High Arctic waters in Brown et al (1975) with the detailed aerial surveys of McLaren (1982) and Renaud, McLaren, and Johnson (1982).

The principal limitation of such work, inevitably, has been its short-term nature. The proponents of industrial projects having to be evaluated are understandably eager to proceed, with a minimal expenditure of time or money in the evaluation process. The usual period allotted to a biological survey is one or two breeding seasons. Yet the years chosen may be aberrant – as, for example, the summers of 1978 and 1979, chosen for the Eastern Arctic Marine Environment Studies (EAMES) baseline surveys, when ice cover in the Canadian Arctic was heavier than in any other summer in the forty-odd years of detailed records (Nettleship 1985). What is more, periods as short as these cover only a fraction of the life spans of many of the most important species. The thick-billed murre (*Uria lomvia*), perhaps the most vulnerable of arctic seabirds, does not even begin to breed until it is five years old, and it has a reproductive life span of the order of ten years (Gaston and Nettleship 1981; Kampp 1982). A 1–2-year survey is bound to produce only a "snapshot" evaluation of the murres' biology, when a "full-length film" is needed to establish the proper baselines for a long-term environmental assessment.

This short-term approach also influences the quality of surveys. The consultants usually maintain a core of personnel and hire additional, temporary staff to meet the requirements of each new contract. Their ornithologists seem especially transient: there is always a pool of bird-watchers to draw on, and also ornithology is considered technically simple and thus gets lower priority than other aspects of environmental surveys. (Few land-based bird-watchers know how to identify and count seabirds at sea or are familiar with the specialized seabird literature.) Even when experienced people have been rehired for a second contract, their work inevitably suffered from lack of continuity, which sometimes appeared as shallowness in the consultants' interpretations of their survey data. Literature reviews tended to be based not on proper library searches but on reviews prepared for other environmental impact assessments. The same phrases – and mistakes – appeared again and again. For example, there is a major movement of thick-billed murres in September out of Hudson Strait and south down the Labrador coast (e.g. Gaston 1980, 1982; Orr and Ward 1982). The migrating birds are swimming, not flying, and would thus be vulnerable to oil spilled at drill-sites on the Labrador Shelf at this season. Yet the initial environmental impact reviews for these sites insisted that the birds avoided Labrador altogether and crossed directly to Greenland instead. An error of such magnitude would have seriously distorted the value of the whole EARP.

The geographical extent of surveys and literature reviews was some-times too narrow. Offshore industrial developments are concerned primarily with the geological and engineering problems associated with specific sites of only a few hectares in area. At first most biological consultants' reports were similarly circumscribed, even where discharged pollutants would inevitably be carried away and would affect highly mobile marine organisms such as mammals, seabirds, and many fish. The absence of these organisms from the site was sometimes cited as evidence of the proposed development's negligible environmental impact. This was claimed in one case, where a large seabird colony was some distance from a mining operation but directly downstream of effluent from the mine tailings, and in other cases, where seabird numbers were low on average but very high during brief migration periods in the spring and fall.

A natural reaction to this limited, site-specific perspective was to make environmental assessments far more complicated than necessary. The distributions and movements of fish, marine mammals, and sea-birds on the Labrador Shelf, for example, cover such a wide area that there is no need to spell them out for every site proposed for offshore hydrocarbon exploration. A general, "regional" statement, with a few site-specific modifications, was usually all that was necessary.

I thought that, on the whole, we in the federal review agencies tended to go too far in the opposite direction and demand a very broad perspective, with no firm guidelines for priorities. The result often read like a grab-bag of information into which every possible environmental fact had been tossed, and some of the priorities were very odd indeed. For historical reasons, the reviews and impact statements for offshore sites discussed geology in considerable detail, physical and chemical oceanography to a certain extent, but biological factors rather poorly. This unevenness reflects the state of our knowledge of these respective disciplines and highlights gaps to be filled.

However, the relative order of environmental interests runs in exactly the opposite direction, with the prime concern being the potential impact of proposed operations on biota. This was sometimes forgotten. I once had to review the contingency plan – for operational use in case of an oil-spill – for a site off Labrador. It discussed in some detail the geological structure and history of the Labrador Marginal Trench – something no oil-spill could possibly affect. Yet it omitted the highly pertinent fact that the whole thick-billed murre population of Hudson Strait passed close to the site in the early fall and that an oil-spill at that time could conceivably eliminate a year-class of young birds and many of their parents. In the absence of this information, the officer in charge of emergency operations might well assume that the biological hazards of a spill were negligible and conclude that there was no urgent need to try to control it.

THE FUTURE OF EARP

EARP is clearly here to stay in Canada, though use of the process has slowed down in recent years, reflecting the slackening pace of exploration of the Canadian offshore region. Georges Bank is the only area off eastern Canada yet to be explored. Possible areas of operation in the "south" have been narrowed down to natural gas reservoirs off central Labrador and around Sable Island and an oil-field near the *Hibernia* well-site on the northeastern Grand Banks. The last two are thought to be commercially viable, though market forces will determine whether they will be exploited. The same applies, further north, to the oil reserves in the Beaufort Sea and the natural gas in the High Arctic islands. When exploration gives way to routine production, there will inevitably be more accidental spills, and probably a decline in safety standards as well (e.g. Anonymous 1977; Bourne 1980). Only then will we be able to judge EARP's effectiveness in predicting, preventing, and coping with accidental discharges, either from offshore production platforms or from tanker traffic, pipelines, and other operations associated with them. We know many of the potential problems, and most of our judgments on immediate environmental risks are based on a reasonably adequate body of background information.

However, our guesses may not be right. If there is any further moral to be drawn at this stage, it must come from our experience of pollution incidents in Canadian waters and elsewhere. Spills from well-sites are usually the result of faulty operating procedures and are therefore preventable, but they have had very few features in common. The Santa Barbara incident (Straughan 1971) was a blow-out of oil and gas, combined with some natural seepage from the seabed, which caused considerable environmental damage. The blow-outs from the *Ekofisk* well-site in the North Sea (Bourne 1977) and from the *Vinland* site off Nova Scotia in 1984 (Environmental Protection Service, Environment Canada, personal communication) released, respectively, crude oil and natural gas which apparently did little harm. The *Ixtoc I* blow-out in the Gulf of Mexico was of both oil and gas, which burned out of control for several months but caused little environmental damage (Waldichuk 1980).

The causes and effects of ship spillages are equally unpredictable. The seemingly negligible spill of seventy tonnes of bunker c from a leaking hatch-cover on *Irving Whale* was just as serious to seabirds as were the thousands of tonnes released from the wrecks of *Arrow*, *Kurdistan*, and *Argo Merchant* (Brown et al 1973; Powers and Rumage 1978; Brown and Johnson 1980). The wrecks of *Arrow*, and of *Argo Merchant* off New England in 1976, apparently resulted from navigational problems that could certainly have been prevented by proper seamanship and maintenance; the break-up of *Kurdistan* was sudden, unpredictable, and not yet explained satisfactorily.

The *Kurdistan* incident posed four separate problems at once, at least two of them completely novel in oil-spill operations (Vandermeulen 1980). There were the ten thousand tonnes of oil remaining in the (salvageable) stern half of the ship; the seven thousand tonnes in the (probably unsalvageable) bow; over a thousand tonnes entrained in pack ice drifting out of Cabot Strait, which soon began to go ashore in eastern Cape Breton; and the expected threat of about seven thousand tonnes released from the centre portion, which drifted undetected in Cabot Strait for three weeks before it went ashore, with devastating results, just as we were congratulating ourselves that the emergency was over. The Regional Environmental Emergency Team (REET) used available environmental information to advise the Canadian Coast Guard on possible courses of action: i.e. tow the stern into harbour in Cape Breton, sink the bow in deep water south of Sable Island, and wait for the remaining oil to come ashore. We had no environmental impact assessments to guide us, but it is unlikely that any assessment could have foreseen all four possibilities, let alone more or less simultaneously.

In other words, we can be certain only of the uncertainty surrounding future oil pollution incidents, whether the spill is from a ship or a rig. The cause of the incident will almost certainly, one way or another, be human error, and the spill itself will probably present some totally new and unforeseen problems. It would therefore be a mistake to be too rigidly elaborate in advance preparations. We must realize, however, that EARP is unlikely to cover every possible short-term contingency and stand by to expect the unexpected.

Assessments become progressively less reliable with a longer view than one or a few months, the time-scale of most oil-spills. This is an inevitable consequence of the short history of environmental studies (and of EARP, in Canada), combined with the short-term nature of many recent surveys. The *Arrow* disaster allowed us to measure the ability of a littoral environment to recover from an oil-pollution event over an eight-year period (e.g. Vandermeulen 1978, 1982). Unfortunately, slow reproductive turnover prohibits such medium- to long-term assessments for the most vulnerable Canadian seabird species (see above). We are even less able to assess the effects of completely new technologies. For example, the Arctic Pilot Project (Northern Component) proposed fifty-six passages a year through Parry Channel and Baffin Bay, with ice-breaker-tankers of 135,000 tonnes, displacement (Anonymous 1980). The obvious hazard is that collision with an iceberg will release fuel oil or set off an explosion of liquid natural gas. Such short-term risks EARP is well equipped to assess. But what will be the long- or even short-term effects on seabirds, marine mammals, and other organisms of the underwater noise generated by this year-round traffic or by the repeated breaking up of the winter ice cover?

We have barely enough information to begin to make realistic assessments of such questions. Yet EARP panels and environmental review agencies are expected to make them, regardless of information gaps. Do we reject or delay a project until more data are available? Or do we give it cautious approval, subject to certain precautions which, we hope, will minimize largely unknown hazards? Neither solution is a very comfortable one. Advice to delay or outright rejection of a project risks being overturned at higher levels of government for political or economic reasons, unless it can be shown that the proposal itself, not just the available data base, is seriously flawed. However, precautions attached to cautious approval may prove impractical in budgetary terms.

The result in either case could be initiation of a major project with few if any environmental checks and balances. I can see no obvious way out of this dilemma.

In modern life economic concerns usually take precedence over environmental ones. The Canadian "environmental conscience" is no longer the burning matter of public concern that it was fifteen years ago. Some, indeed, consider it an unnecessary luxury in today's financial climate, but this is too narrow and pessimistic a point of view. We cannot afford to turn back the clock to the days when a project was judged solely on its economic and engineering feasibility, with no concern for its environmental consequences. Our Environmental Assessment Review Process is not perfect, but it does provide some very necessary safeguards. We must take care that economic considerations do not shift emphasis toward the kind of short-term environmental problems that may be solved relatively cheaply and easily, at the expense of longer-term environmental concerns.

ACKNOWLEDGMENTS

I thank Dr A.J. Erskine for his comments on this manuscript. This paper is associated with the program Studies of Northern Seabirds of the Canadian Wildlife Service, Environment Canada (Report No. 185).

PART TWO

Differing Assessments

CHAPTER FIVE

Industry, Shipping Proposals, and Science

Robert L. Dryden

"You can't tell which way the train went by looking at the track" (anonymous). Nothing expresses better the confusion and frustration experienced by many, if not all, participants in arctic regulatory proceedings. Industry, in particular, can become extremely perplexed by the lack of benchmarks and/or standards with which to judge progress in the environmental and socio-economic review arena. Such frustration, however, is not confined to industry. It extends from the Inuk of Grise Fiord right through to the much-maligned bureaucrat of Ottawa. The regulatory review process for arctic matters resembles closely a carnival "pool of balls," in which young children clamber about in three to four feet of small coloured balls, without ever making any significant progress.

The chapter first summarizes, and it is hoped, interprets, the petroleum industry's perspective on arctic marine transportation and the environmental assessment regime, but many of the problems encountered by industry are common to all participants in the process.

Arctic environmental reviews, without exception, have spread into the realms of politics and socio-economics. As a result, erosion of scientific boundaries and inclusion of the more emotional "people-power" issues have changed the traditional type of review into a grab-bag event more suited to politicians and lawyers than to the environmental scientist. A recent government/industry/university study concluded: "We either improve the scientific rigour of the studies which support the entire process or we run the risk of seeing the concept of environmental impact assessment degenerate into an exercise in public relations and government lobbying" (Beanlands and Duinker 1983). This chapter argues that, from an industrial viewpoint, such degeneration has already seriously eroded the study concept.

This chapter next defuses, if possible, the image of Canada's Arctic

as pristine and as yet unblemished by southern technology. The picture
of a virginal wilderness where people live solely off the land and in
total harmony with nature has been presented to southern audiences
for decades. Almost without exception, the artist painting this image
is also from southern Canada and has just returned from a three-to-
four day summer visit on behalf of his/her newspaper or television-
station employer. The traditional story describes the ignorant and con-
fused aboriginal trying to protect a dog-sled way of life from avaricious
and evil oil barons. In truth, however, industry is no stranger to Canada's Arctic. The dog-team has long disappeared in favour of the much
more efficient snowmobile. Skin kayaks and bows and arrows have
been replaced for many years by the outboard engine and the high-
power scoped rifle. Igloos and even the techniques of making them are
fast becoming only memories held by community elders.

All the symbols of a changing northern way of life are based on the
growing use of petroleum products for fuel and construction material.
At the present time, there are many arctic industrial mega-projects. The
oil and gas industry has been very visible in both the Beaufort Sea and
at Norman Wells. Major mining projects have been a fact of northern
life for many years in Coppermine, Nanisivik on Baffin Island, and
Polaris on Little Cornwallis Island. Additionally, Panarctic has been
conducting a major exploration program for almost twenty years in
the Melville-Lougheed–King Christian Island areas. The north has also
been intensively introduced to further oil and gas activities through the
past ten years of the Arctic Pilot Project, Polar Gas Project, and work
in Lancaster Sound.

Of these companies and projects, Gulf, Dome, and Esso are good
examples of how industry is already interacting with northern communities. In the first six months of 1982, Gulf injected $4.68 million
into the local Beaufort Sea area by way of thirty-seven separate supply
and service contracts (Gulf 1982). Dome has been equally active in
promoting local business. In 1981, it purchased in excess of $24 million
worth of goods and services from local businessmen in the Beaufort
area. Their total business expenditures in Yukon and Northwest Territories that same year were approximately $35 million (CPA 1982). Esso,
for its part, has established local business development programs in the
Beaufort Sea area and in Norman Wells. The Norman Wells pipeline
project was anticipated to provide $40 million of supply and service
contracts to northern business (CPA 1982).

This chapter lastly presents industry's side in response to the oft-
stated criticism that its only interest in the environmental assessment
process is the final project licence and approvals. Industry is accused of

doing the absolute minimum amount of work while hiding or misrep-
resenting anything having negative environmental implications. In the
past some of these concerns might have been justified. In a few instances,
with some of the less diversified companies, such claims might yet have
substance. In total, however, the oil and gas industry has learned from
past mistakes that there is little to gain and much to lose in attempting
to cheat in the process of regulatory review. Entirely too much is at
stake to risk not using the best information available.

Perhaps part of such a change can be attributed to the northern "boom"
cycle of the 1970s, when oil and gas companies finally understood that
the system was too big to ignore. During this period companies bla-
tantly raided government agencies, in order to procure the cream of
environmental scientists. Now that those frantic days have passed and
the industry oversupply of environmentalists has returned to government or to private consulting, industry retains a new level of environmental competence and an overall improved understanding and
appreciation of environmental practices and philosophies.

ARCTIC MARINE TRANSPORT: HISTORY AND EXTENT

As a specific extension of the previous discussion of the arctic industry
in general, a brief outline of the historical and current usage of arctic
waterways is in order. The increasing dependence of northern residents
on marine transportation perhaps highlights the anomalies in special
interest groups' condemnation of northern hydrocarbon shipping
proposals.

The first wave of sailors in the Canadian Arctic involved the Norse-
men of the eleventh century in search of areas suitable for settlement.
After the Norsemen lost interest in the Baffin Island area, there was no
European contact with the Canadian Arctic until the sixteenth century
(Hocking and Anderson 1980). At that time the European Renaissance
led to a dramatic expansion of European influence throughout the world.
The British, during this period, attempted to find a shorter route to
the silks, spices, and other wealth of Asia. The first searches for the
Northwest Passage were for commercial or industrial purposes and
involved such famous sailors as William Baffin and Martin Frobisher.

The next phase in Canadian arctic marine voyages was British arctic
exploration during the nineteenth century. The search for the Northwest Passage was again the primary objective, but not to find a shorter
route to the Orient. Rather, national pride, scientific curiosity, and
questions of sovereignty motivated the British Admiralty to undertake
a half-century of exploration. The 1845 Franklin Expedition, by its

failure and the many follow-up search and rescue expeditions, eventually led to recognition and acceptance that the Northwest Passage did indeed exist.

While the scientific voyages of discovery were occurring, the British whaling industry was also increasing its activity. Whale oil, baleen, and jawbones of arctic whales commanded a high price in nineteenth-century Europe. With the further opening of Canada's arctic seaways, whalers were able to enter previously untapped hunting grounds. The peak of whaling activity occurred probably with the eighty-seven ships that sailed into Baffin Bay in 1830. By 1840, through loss of profits and loss of ships to ice, there remained but twenty-seven whaling vessels in the Canadian Arctic. The second half of the nineteenth century saw the transition from sail to steam-powered vessels, which could move through previously impassable ice. Greater range capabilities also allowed steam vessels to expand into previously unexplored bays and inlets such as Cumberland Sound.

By the early twentieth century, the whaling industry was near collapse, because of declining whale populations and the development of commercial substitutes for such items as whalebone. Arctic shipping, therefore, gradually evolved from 1900 to 1940 into a program to assert Canadian sovereignty and resettle communities. Fur hunting and mineral exploration replaced the search for whales.

More recently northern shipping has switched to the more traditional function of resupplying northern communities. Foodstuffs, fuel (both bulk for heating and containerized for skidoos, boats, and so on) oil lubricants, medicines, building materials, household furnishing, and many other products depend on arctic shipping. The following information and statistics (Canadian Coast Guard 1983) show that arctic shipping and increasing use of ice-breakers and ice-strengthened freighters and tankers are crucial to northern communities. (They exclude current, rather extensive marine resupply associated with oil and gas activities in the Beaufort Sea area.)

Four sea routes have traditionally been used for arctic resupply: the eastern Arctic route, the District of Keewatin route, the Mackenzie route, and the Pacific route, through the Bering Strait. The eastern Arctic route originates in eastern Canada and Ogdensburg, New York, and passes through Hudson Strait into Hudson Bay and Foxe Basin and on to Lancaster Sound and northward. The District of Keewatin is served by a tug-barge operation out of Churchill. The Mackenzie route extends from Hay River to Tuktoyaktuk. The Pacific route provides for the annual transit of tugs or barges and ice-breaker-escorted cargo from west coast ports to the Alaskan and Canadian Beaufort. In addition, on the east coast, the ice-strengthened MV *Arctic* resupplies on a Europe-to–Lancaster Sound routing across the North Atlantic. Appendix A identifies and categorizes the Coast Guard documentation of 176 transits by resupply vessels that crossed 60°N into and out of the Arctic during 1980. Vessels included tankers carrying bulk POL (petroleum, oil, and lubricants) and general cargo vessels carrying all other items, including drummed fuel. The size of vessels involved ranged from 500 tonnes to 20,000 tonnes and up. These sizeable numbers of commercial cargo vessels crossing into and out of Canada's Arctic every year do not include the inland Mackenzie and Keewatin routes. The Coast Guard also predicts that, based strictly on northern population growth, per capita consumption rates, and the increasing standard of living, arctic resupply vessel transits will increase significantly by 1990.

The Arctic has seen also increasing use of ice-infested channels for commercial vessels involved with mineral extraction. With ore-concentrate shipping from such places as Nanisivik on Baffin Island, Polaris on Little Cornwallis Island, and the Asbestos Hill mine at Deception Bay, the number of large ore-carrying freighters is quickly becoming a feature of the north. In 1980 the number of one-way arctic transits of mineral carriers already stood at fifty-six. The Coast Guard predicts that that number will increase to seventy-two by 1990. At a slightly more southerly location, grain is shipped through the port of Churchill, Manitoba, through Hudson Bay, and east to Europe. In 1980, seventy-four transits of grain-carrying ships were made to and from Churchill (Canadian Coast Guard 1980).

One of the more ambitious arctic shipping programs involves the ice-breaking service provided by the federal government. In recent years, demands for ice-breaking services have expanded from traditional support of navigation to include control of vessel-source pollution in ice-infested waters, enforcement of the Arctic Waters Pollution Prevention Act, and assistance in search and rescue, hydrography, seismology, and research. Moreover, Coast Guard ships are used to deliver cargo to sites not serviced by commercial carriers, such as the Eureka weather station and other non-settlement scientific sites. Coast Guard ice-breaking services are provided by vessels in several power ranges, including the lighter ice-breakers of 4,000–6,000 hp such as *Bernier*, more capable vessels of 13,000–16,000 hp such as *Sir John Franklin* and *Pierre Radisson*, and the larger, 24,000-hp *Louis S. St. Laurent*. The ice of Arctic Canada has also been well researched and tested by industrial work undertaken by SS *Manhattan*, MV *Kigoriak*, and MV *Arctic*.

In addition, various federal, territorial, and provincial agencies also use the northern shipping lanes to provide social-welfare, educational,

medical, legal, and administrative services to isolated northern communities. The Canadian Forces' presence in the Arctic also necessitates
resupply vessels to operate and maintain their observation posts.

None of the vessels operating in and out of the Arctic in open-water
"seasons" or beyond has ever undergone environmental impact assessment. Even given that pollution is normally associated only with
hydrocarbon transport in ice-covered waters, there are many other
issues, such as underwater noise, discharge ballast, dredging of northern
terminals, disruption of ice dynamics (by Coast Guard ice-breakers, by
MV *Arctic*), and disturbance of seabird colonies.

LEGISLATION

The process of environmental and socio-economic review for oil and
gas industry proposals involving marine transportation is complex and
rife with duplication. To quote a particularly relevant source: "Environmental issues in the Arctic fall within the jurisdiction of the Department of Transport, Fisheries and Oceans, Environment, Indian
Affairs and Northern Development, as well as the Government of the
Northwest Territories. Each of these agencies appears to have powers
that, if exercised, could restrict passage of ships through the Northwest
Passage" (FEARO 1980). Another study commissioned by Transport Canada
discussed industry's uncertainty because of the diversity and complexity
of environmental requirements; it concluded: "A major factor in the
uncertainty is the proliferation of legislation and agencies involved in
regulation of Arctic marine operations" (Albery et al 1978).

The process of review and control for projects proposed for the
Canadian Arctic theoretically rests with the federal Department of Indian Affairs and Northern Development (DIAND), which co-ordinates
and oversees the activities and programs of all federal departments in
northern Canada. In brief, DIAND is the "key actor" in all aspects of
northern affairs, including resource development and environmental
management. It is not, however, the only department with responsibilities that affect project development. As indicated in the FEARO 1980
quotation, Environment Canada, Fisheries and Oceans, Transport Canada, and the government of the Northwest Territories also have a say
about the environmental acceptability of development proposals. In
addition, the Canada Oil and Gas Lands Administration and the National Energy Board are directly involved, and both agencies have recently expanded their northern mandates to include environmental and
socio-economic issues.

With the bewildering array of regulatory legislation and requirements
and the functional overlap of activities, one agency (i.e. DIAND) can

hardly act as a co-ordinating body. For instance, a report prepared for the Canadian Arctic Resources Committee identified five subcommittees of the Advisory Committee on Northern Development (ACND) plus eighteen non-ACND committees in six different federal departments, all concerned with arctic marine transport and northern development (Dosman 1979). Naturally, all committees and subcommittees establish additional working groups and task forces to deal with specific issues.

The remainder of this section attempts to identify and briefly summarize the most relevant environmentally related statutory and guideline requirements facing an arctic marine transport proposal. The list is far from exhaustive and does not include all the twenty-eight statutory requirements identified by Albery (Albery et al 1978).

Statutes and Guidelines

Canada Shipping Act: The Canada Shipping Act applies to all Canadian waters north of 60° N not within shipping safety control zones established under the Arctic Waters Pollution Prevention Act, out to the 200-nautical-mile limit of Canadian fishing zones. It establishes detailed ship construction standards and operating procedures. Standards are set also for discharge by ships of oil, garbage, air contaminants, and other, specified polluting substances.

It is a summary-conviction offence for any ship to discharge a "pollutant" as defined in the act and regulations. Ships' masters are required to report pollutant discharges to pollution prevention officers appointed under the act. These officers have wide powers to inspect and compel the releasing of information. They are authorized to make orders respecting ship movement and clean-up of pollutants. Under the act, owners of bulk-pollutant carriers and certain cargo owners are made civilly liable for damage to life or property and for clean-up costs resulting from discharge of a pollutant. Liability is strict; it does not require proof of fault or negligence.

Fisheries Act: The Fisheries Act basically sets water pollution standards. It legislates fines of up to $100,000 per day on conviction for depositing any harmful substance into waters frequented by fish. It is also an offence to alter or destroy fish habitat. Regulations developed under the act identify contaminant standards. Like the Canada Shipping Act and the Arctic Waters Pollution Prevention Act, this act makes persons who own, control, or cause deposit of deleterious substances civilly liable for the costs of control measures and clean-up. Again liability is strict. The minister of fisheries and oceans is also empowered to request plans, specifications, and other information concerning any project that might result in the deposit of harmful substances. An assessment of such information could result in the minister ordering modifications or prohibition of a project.

Arctic Waters Pollution Prevention Act: The AWPPA controls or prevents pollution of arctic waters in three ways. First, it makes deposit of waste an offence punishable on summary conviction by a fine of up to $100,000 for each day the offence is continued. Second, it makes ship and cargo owners civilly liable without proof of fault or negligence for all loss and damage suffered by persons and for clean-up costs. Third, it establishes shipping safety control zones based on navigational and ice conditions. The act also regulates the discharge of waste from shore-based marine terminals and facilities.

Navigable Waters Protection Act: This act provides that no work (e.g. dock, pier, or wharf) shall be built or placed in navigable water, without approval of the minister of transport.

TERMPOL Code: The Code of Recommended Standards for the Prevention of Pollution in Marine Terminal Systems (TERMPOL) is a voluntary interdepartmental process designed to facilitate assessment of navigational and environmental risks associated with the location and operation of marine terminals. The code is aimed primarily at oil pollution risks.

Canadian Environmental Protection Act: This act, incorporating the previous Ocean Dumping Control Act, prohibits the deliberate dumping of any substance at sea, except under permit from the minister of the environment. The act designates particular substances as prohibited or restricted.

Territorial Lands Act: All operations on crown land in Northwest Territories, other than hunting, trapping, and fishing, require a land use permit. Marine terminal facilities fall under this system. Environmental and operating conditions may be imposed with the granting of a permit.

THE REVIEW PROCESS

As can be seen from the previous section, environmentally related aspects of arctic shipping are well legislated. How do regulatory agencies determine whether a development proposal satisfies the various statutes and regulations? This is where the environmental review process is applied. This is also where the government's role of control and co-ordination begins to waver.

EARP

The Environmental Assessment and Review Process (EARP) was established by a 1973 cabinet decision and strengthened by order-in-council in June 1984. This process was intended as a "one-window" approach to the environmental review mandate of all federal departments having responsibility for environmental issues. All projects or proposals being initiated by the federal government, having federal funding, or being on federal property fall within the compass of EARP. All projects located on Canada Lands, north of 60°N, must therefore be considered by EARP.

The first step in the process is a project self-examination for potentially significant environmental effects. If such potential does not exist, the project may theoritically proceed to implementation. Because emphasis is placed on northern environmental protection by the the government, however, almost any oil- and gas-related proposal will contain at least a few "significant" concerns. Assuming such a finding, project proponents must next prepare an initial environmental evaluation (IEE) which, in turn, leads to a project referral for a formal review under EARP. Formal review involves a much more detailed environmental impact statement (EIS) and a series of technical, scientific, and public reviews of all anticipated environmental and related social effects of the project.

EARP does not have a legislative basis, as it was created by cabinet. It is therefore extremely broad-based and wide-ranging. Its mandate is also very flexible and can be amended simply through cabinet decision. EARP tackles each and every environmental and socio-economic issue relevant to a proposal. It was intended to, and does, incorporate the scientific input of all agencies and public interest groups. It is the melting-pot and focus for the entire gambit of environmental thinking. In theory, once a proposal has been through EARP, then it should require no further environmental or social examination. Such, however, is not the reality.

TERMPOL

As stated earlier, TERMPOL review is intended to identify and measure navigational and environmental risks associated with the siting and subsequent operation of marine terminals. Although participation is voluntary, the information provided to the Coast Guard is the basis for special regulations and/or operating conditions. Misinformation could well result in onerous and unnecessary restrictions.

If a full TERMPOL review is initiated, a co-ordinating committee is established. The committee establishes several steering committees and subcommittees. Representation on steering committees comes from a number of federal departments, including Fisheries and Oceans, Public Works, Environment Canada, Transport Canada, and DIAND. Terms of reference can be very broad and may require information ranging from terminal design and systems operations to contingency/emergency response planning and environmental effects and mitigation.

National Energy Board

The National Energy Board (NEB) is an independent agency mandated to analyse Canadian energy projects. Its governing statute limits the board's regulatory authority to granting energy export licences, approving interprovincial or international pipeline facilities, and setting rates, tariffs, and tolls for such pipelines (Lucas 1979). It has no specific authority to regulate the marine transport of hydrocarbons, unless export is involved, or to regulate matters concerning the environment or the socio-economic effects of a project.

The act does grant, however, power and responsibility (under section 11) to inquire into all energy matters within the federal government's legislative jurisdiction or the public interest. Exercise of this broad and unbounded authority depends either on the board's own initiative or on the direction of the minister of energy, mines, and resources. The question of competence and experience with northern issues is not a factor in the board's ability to hold public hearings to review issues such as potential environmental and socio-economic effects, Canadian content, or economic costs and benefits of a proposed undertaking. As with the Arctic Pilot Project, details of which are reviewed later, the board does not even have to solicit input from functional environmental regulatory agencies, such as Fisheries and Oceans and Environment Canada.

The format for the NEB review involves an application (e.g. for a permit to export) to the board. Following distribution and review of filed material to all interested members of the public, a series of quasi-judicial public hearings is convened. This is the only review process that requires legal counsel for the presentation and cross-examination of project evidence. Following the hearing process, the board reports its findings to the minister.

Shipping Control Authority

As a direct result of the 1980 EARP report into the Arctic Pilot Project, the federal government established a Department of Transport Shipping Control Authority, with a mandate to "monitor and manage Arctic shipping and ship routes in the interests of ship safety, the efficient movement of ships and the protection and preservation of the Arctic environment" (Canadian Coast Guard 1983). The control authority established the Environmental Advisory Committee on Arctic Marine Transportation (EACAMT), to include representatives from Fisheries and Oceans, Environment Canada, Transport Canada, National Defence, DIAND, the government of Northwest Territories, and members of the Inuit community. The committee also included representatives of industry (Dome Petroleum and the Arctic Pilot Project).

EACAMT assesses environmental information relating to arctic marine proposals in order to: define data gaps, recommend areas of study, establish environmental monitoring systems, and minimize disruption of the environment. It can recommend measures to minimize disruption of other human activities such as hunting, fishing, and trapping.

The initial organizational structure proposed a working-group/task-force substructure divided into five or six issue areas, including guideline development, environmental research, monitoring, contingency plans, and the review of individual shipping proposals (Nicholls 1986).

Northern Land Use Planning

In 1979, as a result of the potential for hydrocarbon initiatives in Lancaster Sound, the government of Canada undertook a major interdepartmental analysis, the Lancaster Sound Regional Study. This exercise was designed to review all existing knowledge of the physical and biological environment, the human/socio-economic status, and the range of current land uses for the study area, in order to produce a long-range plan of land use options and recommendations. What actually evolved after almost four years of study and public hearings was little more than a series of platitudes and extremely conservative and general options, presented as a government Green Paper. Study reports were made public, in 1982 and 1983 (Jacobs 1981; Dirschl 1982; Jacobs and Palluq 1983).

More relevant perhaps are the after-effects of government becoming involved in land use planning for northern Canada. In October 1982, DIAND issued a draft policy, "Land Use Planning in Northern Canada" (DIAND 1982). This program and/or policy was not intended to replace any existing planning mechanisms, rather it is designed to complement them" (DIAND 1982). The plan itself identified an extremely cumbersome process – regional planning areas, northern land use planning committees, northern land use planning commissions, and area planning teams – to produce land use plans for six regions and fifty sub-regions of Canada's Arctic. The draft plan stated: "Northern Land Use Plans will encompass all interacting components of the environment: land, including offshore submerged lands; fresh and marine waters; sea ice; and renewable and non-renewable resources. The social and economic environment will also be included to the extent appropriate for land use planning" (DIAND 1982).

Although the process is only now in the initial planning phase, the apparent complexities of the proposed system do not bode well for future project proponents. The federal-territorial agreement on land-use planning for Northwest Territories does not provide clear guidance on the relation of land use planning to other planning processes and does not change the jurisdictional mandates of powerful government agencies such as the Canada Oil and Gas Lands Administration (Richardson 1987).

Special Interest Groups

In 1984 the Committee for Original Peoples' Entitlement (COPE) concluded an agreement with the federal government that gives the Inuvialuit of northern Canada surface and sub-surface rights to specific lands of the western Arctic – in the first settlement of several native land claims now being negotiated in the region. The Inuvialuit Final Agreement (DIAND 1984) provides for an environmental impact screening and review process and for a wildlife compensation process. These without doubt loom as the greatest threat yet to industry in the north. The powers of the agreement apply not just to the five thousand square miles of land having mineral and granular rights attached, but also to all lands and waters within the overall Inuvialuit Settlement Region. The total settlement area is thirty-five thousand square miles of lands, including the beds of rivers, lakes, and oceans. The entire Mackenzie Delta and all waters north of the delta are included.

The Impact Screening Committee and the Environmental Impact Review Board established to oversee the region each have seven members – three appointed by Canada, three by the Inuvialuit, and a chair appointed by Canada, but only with the consent of the Inuvialuit. This is the first permanent environmental review process with permanent members identified for a site-specific geographical area in Canada. The EARP process, in contrast, appoints a panel on a project-by-project basis, with members' tenure lasting only until conclusion of the review. With the Canada/Inuit system, the temptations may be overwhelming for review members to look for additional work in order to justify their existence.

Additionally disturbing is the tremendous duplication of effort in an

Inuit review and the many other northern environmental review arenas. Both the initial impact screening and the more formal impact review are close copies of the federal EARP. Under the Inuit system, the initiator of a specific development proposal (normally DIAND for northern projects) no longer initially self-examines a project's environmental significance. Instead, this is done by a committee structure entirely removed from first-hand knowledge of project particulars. A grim forewarning of the impending duplication is provided in section 11 (32) of the agreement: "For greater certainty, nothing in this section restricts the power or obligation of the Government to carry out environmental impact assessment and review under the laws and policies of Canada" (DIAND 1984).

The other major area of concern for industry, and all believers in fair play, is compensation for damage to wildlife, its habitat, and Inuit harvesting. That there should be payment and mitigation for actual and demonstrable damage to wildlife cannot be disputed. However, the agreement attempts to define the traditionally vague and complex question of "future" losses of wildlife and harvests. Researchers, lawyers, and soothsayers have all struggled with the definition and prediction of future events. The primary difficulty is assignment of cause and effect. A hypothetical situation could involve a noticeable decline in wildlife and harvests twenty years after an industry enters a region. How much of this decline will be attributable to industry and how much to other factors, such as natural variability (e.g. climatic changes and natural migration of animal populations from region to region), technological evolution (high-powered rifles and snowmobiles v. harpoons and kayaks), and the unpredictability of human nature? It would seem impossible to predict how attracted people will be to the hardships of living off the land.

For the reasons just given, there appears justification for industrial qualms concerning the resolution of "future harvest loss" claims. The following sections, taken from the wildlife compensation portion of the Final Agreement (DIAND 1984), demonstrate the seriousness of the issue:

13.(7) Every proposed development of consequence to the Inuvialuit Settlement Region that is likely to cause a negative environmental impact shall be screened by the Screening Committee to determine whether the development could have a significant negative impact on present or *future* wildlife harvesting ...

13.(11) Where ... a proposal is referred to the Review Board, it shall, on the basis of the evidence and information before it, recommend to the government authority empowered to approve the proposed development, an estimate of the potential liability of the developer, *determined on a worst case scenario*, taking into consideration the balance between economic factors, *including the ability of the developer to pay*, and environmental factors ...

13.(15) Where it is established that actual loss or *future harvest loss* was caused by development, the liability of the developer shall be absolute and he shall be liable without proof of fault or negligence for compensation to the Inuvialuit, and for the cost of mitigative and remedial measures ... (emphasis added).

With this type of major contradiction, such agreements could reverse recent trends toward greater goodwill, trust, and openness between industry, government, and northern residents.

Discretionary Review

In addition to the visible regulatory review processes just discussed, ministers and cabinet have tacit but powerful and discretionary powers to order additional reviews and inquiries into matters not normally within any specific mandate. The NEB's general inquiry into all aspects of the Arctic Pilot Project is an example. The federal cabinet can also, at any time, establish a commission of inquiry, under either the Territorial Lands Act or the Inquiries Act, to review and report on any aspect of each and every oil and gas development – as with the Mackenzie Valley Pipeline Hearings and the Alaska Highway Pipeline Inquiry (Hunt and Lucas 1980).

A TEST CASE

Since the early 1970s, both the federal government and Canada's scientific community have recognized and voiced the need to improve the country's understanding and use of arctic waters. In 1970, the minister of DIAND identified Canada's principal northern interests as security, economic development, and preservation of the northern ecological balance (DIAND 1970). In his speech, Jean Chretien stated that the Canadian government wanted arctic waters to be opened to commercial traffic. The federal Arctic Waters Pollution Prevention Act (AWPPA) legislated shipping zones and mandatory standards for vessel operations in the Arctic. This act was also a recognition that Canada, with one of the largest arctic coastlines, was in severe danger of falling forever behind, both in arctic transportation technology and in safe exploitation of northern natural resources.

In 1971, the Science Council of Canada recommended immediate attention to arctic navigation, in order better to develop and use arctic resources. In 1975 and 1979 the council again stressed the need for enhanced arctic transport capabilities, particularly through smaller-scale pilot projects that could be monitored and evaluated (Science Council of Canada 1975, 1979).

In 1977, a seminar, "Natural Gas from the Arctic by Marine Mode," was sponsored jointly by the Atlantic Provinces Economic Council and the Science Council of Canada. Participants at this seminar and in a follow-up report condemned the federal government for neglecting northern development and urged immediate action to strengthen Canadian technological, scientific, and industrial capability in the Arctic (Science Council of Canada 1977).

In spite of many declarations and commitments, there were by 1977 few, if any, indications that the science of arctic marine transport was making headway. In 1977, however, a consortium of industrial interests proposed to open the Northwest Passage to year-round commercial shipping. The Arctic Pilot Project (APP) was to be a small-scale commercial undertaking with a very large-scale research and development component. It was patterned along the lines of government and scientific requirements for arctic Canada. It was also, unfortunately for its proponents, a perfect vehicle on which the languishing review process for northern marine transport and the environment could sharpen its teeth. The APP unwittingly became the forerunner for future northern marine transport proposals.

The rigorous examination afforded the APP demonstrates the real-life application of the environmental review process. Brief mention will also be made of other northern development proposals, including the Lancaster Sound Project and the Beaufort Sea program.

I do not intend to argue the merits of environmental or scientific issues raised. Various issues will be identified, in order to point out the duplication, overlap, and redundancy that occurs from one review process to the next.

The APP was, until its official demise in 1984, conceived and managed by a consortium consisting of Petro-Canada, Dome Petroleum, Nova An Alberta Corporation, and Melville Shipping. The APP proposed to produce and deliver natural gas from Panarctic Oil Ltd's Drake Point gas discoveries, on the northern coast of Melville Island, in the Queen Elizabeth Islands area of the High Arctic. Specifically, the gas was to be transported a short distance by overland pipeline to Bridport Inlet, on the southern coast of Melville Island. There it was to be liquefied and loaded aboard specially designed ice-breaking bulk-fuel carriers for shipment to southern markets. At the southern terminal, proposed for either Quebec or Nova Scotia, the liquefied natural gas (LNG) was to be regasified for distribution by pipeline to commercial North American markets.

This project envisioned development of a commercial, navigable, year-round marine transport route through the Northwest Passage. The route was to be approximately 5,600 km in length and would traverse eastern Viscount Melville Sound, Barrow Strait, Lancaster Sound, and northern Baffin Bay. From Baffin Bay the route would swing south through Davis Strait and the Labrador Sea to the southeast coast of Canada.

The APP was to involve construction and operation of two arctic class-7 ice-breaking vessels, 395 m long, 50 m wide, and with an LNG cargo capacity of 140,000 cu m. Each vessel would be powered by two 75,000-hp combined gas-turbine and steam-turbine propulsion systems and would carry state-of-the-art ice-management and integrated navigational equipment. Each vessel would make about fifteen round trips annually (APP 1982). The fifteen round trips per year, when averaged out over open-water and ice-covered seasons, would entail each ship passing any specific spot along the route approximately once every week and a half to two weeks.

Between 1977 – when the concept was first presented to government – and 1983 – when, for all intents and purposes, it was withdrawn from active consideration – the APP went through one of the longest and most intensive, costly, and circuitous environmental reviews that any project has had to endure. We will look at several distinct processes: environmental assessments by Environment Canada, by Quebec, and by Nova Scotia and the federal government, the TERMPOL process of Transport Canada, and reviews by a federal interdepartmental working group, by a Denmark-Canada working group, and by the National Energy Board. During the six years of review, the APP spent over $60 million in development and review of its proposal (Bruchet and Robertson 1983).

EARP (1977–80)

The federal EARP was conducted by the Department of Environment. It was first initiated in November 1977, when Petro-Canada and DIAND jointly referred the project to Environment Canada (FEARO 1980). From initiation until the Environmental Assessment Panel Report of October 1980, the APP was required to prepare and submit a vast number of environmental impact statements (EISs), supplementary information documents, and environmental study reports. The panel also requested and reviewed all related social and economic factors associated with the APP; its mandate had been expanded in 1979 "to include the examination of the political socio-economic implications of the project" (FEARO 1980).

The federal EARP was both professional and comprehensive. The series of community meetings and formal hearings held in northern communities perhaps indicated the intensity of the project assessment. Governmental participants included federal DIAND, Energy, Mines and Resources, Environment, Fisheries and Oceans, and Transport and the NWT government. Non-government organizations included the Baffin Regional Council, the Baffin Regional Inuit Association, the Canadian Arctic Resources Committee, the Inuit Tapirisat of Canada, and representatives of Arctic Bay, Grise Fiord, Pond Inlet, and Resolute. Many non-aligned individuals also made presentations.

Throughout the EARP review, a number of issues were raised and reviewed time and time again. These same issues, although responded to and satisfied within the panel report, were to resurface and be re-reviewed throughout all the other APP review processes. These issues included:

- concern that allowing passage of the two APP ships would result in hundreds of ships in the area and thousands of annual transits;
- the ship routing corridor passing too close to the coast of Greenland (Greenland's initial environmental concerns led the APP to reroute its carriers out of coastal Greenland waters and into a corridor that passed, more or less, down the center of Baffin Bay and Davis Strait);
- northern terminal issues, including ballast water discharge, ice-management disruptions to Bridport Inlet ecology, dredging, and shoreline disturbances of muskoxen and caribou;
- capability to predict and monitor sea-ice and weather conditions;
- disruption of sea-ice patterns and dynamics as a result of ice-breaking;
- safety with respect to explosions and/or fire resulting from LNG spill;
- ocean pollution as a result of LNG or diesel-fuel spills and containment and recovery capability;
- ice-breaking disruptions to ringed seal denning, entrapment of whales in ship tracks, Inuit hunting disruptions due to open water in ship tracks, and caribou migrations across Parry Channel;
- damage and disruption to marine mammals caused by ship noise;
- damage and disruption to seabirds caused by ship traffic and LNG spills;
- social factors including land claims, local employment, and the boom/bust cycle of development.

These points are representative of the issues most frequently raised by intervenors and demonstrate the lack of co-ordination in the environmental review process in the north.

The overall conclusion of FEARO's three-year review was: "The Environmental Assessment Panel has reviewed the northern component of the Arctic Pilot Project and has found the project to be environmentally acceptable subject to certain conditions" (FEARO 1980). Further, "The Panel is able to endorse, subject to certain conditions, limited

shipping on a year-round basis as proposed by the Arctic Pilot Project"
(FEARO 1980).

Examples of "conditions" included establishment of a monitoring
and control mechanism and environmental advisory committee for arctic shipping, plus continuing research into marine mammals and the
effects of underwater sound. The first was fulfilled by establishment of
the Coast Guard Shipping Control Authority and Environmental Advisory Committee. The second and other similar conditions were set
toward fulfilment by the APP's commitment to a $200-million APP Re-
search and Development Program.

Quebec (1978–81)

Both the Gros Cacouna site in Quebec and Melford Point site in Nova
Scotia were acceptable to the project from a technical viewpoint, but
selection and designation of the southern terminal were left in the hands
of government, largely because of political factors. The powers-that-
be decided on formal site-selection reviews for both locations.

The government of Quebec conducted an internal review of the
project from December 1978 to October 1980. This review, which
included public hearings, was intended to select the optimal terminal
site in Quebec. Available marine technology and the environment were
the primary criteria for selection. Following approval and announcement of the Gros Cacouna location, a second series of Quebec public
hearings was held during January and February 1981. The Canadian
Coast Guard's TERMPOL process played a significant role in the public
hearings.

In March 1981, after two and a half years of review, the Bureau
d'audiences publiques (Quebec's counterpart to the EARP) reported to
the Quebec Ministry of the Environment that it found the project "as
a whole and as submitted to be acceptable biophysically, socially, economically and in terms of safety" (NEB 1982b).

Nova Scotia and Canada (1979–81)

In Nova Scotia, a joint federal-provincial Environmental Assessment
Panel was formed to review the APP shipping route south of 60°N. Again,
after submitting mountains of project and environmental study documentation, including a separate EIS for Melford Point, the APP was once
more invited to participate in a series of public meetings and hearings.
Transport Canada's TERMPOL process was again implemented. After two
years of review, the Canada–Nova Scotia Assessment Panel found in
August 1981 that "the Project can be carried out in a satisfactory manner
with regard to both safety and environmental aspects" (NEB 1982a).

TERMPOL (1979–81)

The APP, in applying to Transport Canada for permits under the Navigable Waters Protection Act and the Arctic Waters Pollution Prevention Act, also presented itself in January 1979 for a full and formal TERMPOL Review. Over the next three years, Transport Canada undertook a long and and very complex committee-subcommittee assessment of the project. Three independent, interdepartmental TERMPOL reviews of the three proposed terminals (i.e., Bridport Inlet, Melford Point, and Gros Cacouna) were conducted. Each review was managed by a separate co-ordinating committee and a steering committee. Federal departments participating on the steering committees once again included Transport Canada, Environment Canada, Fisheries and Oceans, Public Works, and, for the northern terminal, DIAND and the National Research Council. In addition, each terminal review required subcommittees on ship accessibility, ship terminal infrastructure, socio-economic/environmental risk analysis, and design and ship stability.

The TERMPOL process also reviewed in depth the entire north-to-south marine routing proposal, marine pollution and safety, and the social and environmental risks associated with shipping LNG. The issues identified earlier for the federal EARP also applied to this review.

In October 1981, after almost three full years of examination, the Canadian Coast Guard released its report. The executive summary made the following points:

Based on information provided by the proponent, each of the three proposed sites appear to be acceptable in terms of ship safety, subject to the various Committee observations and recommendations.

On the question of safety and marine pollution prevention which are primary Coast Guard concerns the current preliminary review results in an opinion that project plans to date are acceptable.

The proponent's seven volume submission, selected consultant's reports and demonstrated willingness to meet with TERMPOL Coordinating Committee members is acknowledged. This process has, in the Coast Guard view, been a positive approach to the TERMPOL Code assessment procedure. (Transport Canada 1981).

Federal Working Group (1982–3)

The final report of the federal EARP strongly recommended that "selection of the ship routing should involve the integration of physical factors and biological factors so as to minimize adverse impacts on wildlife" (FEARO 1980). The APP, in response, developed a three-volume *Arctic Pilot Project Integrated Route Analysis* (IRA) (APP 1981a). Certainly the most comprehensive biophysical description ever produced for off-shore arctic waters, the IRA reviewed and integrated biological data (mammal/bird populations, distributions, etc) with physical data (meteorology, oceanography, ice dynamics, etc) and resource-use information (harvesting statistics, etc), in an attempt to identify the safest and most environmentally preferred shipping corridor.

While the route analysis was being finalized, the federal government was establishing the Shipping Control Authority and the environmental advisory committee. The latter formed a special project environmental working group, to review the APP route analysis. The working group included representatives from Fisheries and Oceans, Transport Canada, Environment Canada (Atmospheric Environment Service and Canadian Wildlife Service), DIAND, government of the Northwest Territories, Dome Petroleum, and the APP.

The working group began its assessment in early 1982 and examined such issues as underwater noise, vessel safety and design, ice-breaking effects on the physical environment, marine mammal distributions, Inuit hunting, ship-track crossings, seabird colonies, whales in ship tracks, ringed seal denning disturbances, corridor width and route selection method, and contingency planning. Its report, after more than a year of study, concluded: "The consensus of the Working Group is that, based on present knowledge, any environmental impact resulting from existing traffic and the APP operating in the corridor will be within acceptable limits ... In conclusion the EACAMT Working Group recommends that the Control Authority take the necessary steps to support and provide services to shipping on a year-round basis in the corridor as proposed in the Arctic Pilot Project Integrated Route Analysis" (Canadian Coast Guard 1983).

Denmark-Canada Working Group (1979–84)

The project review process established between DIAND, the APP, and Denmark's Ministry for Greenland was not a result of a regulatory requirement. It occurred because of concerns raised by Danish officials that APP ships would pass so close to the coast of Greenland that they would disrupt traditional Inuit offshore harvests of marine mammals. Several meetings between government officials and the APP during 1978 and 1979 led to a joint APP/Denmark/Canada environmental studies program and review group. The first meeting of the Arctic Pilot Project Working Group (APPWG) was held in late 1979. Its mandate included review and provision of advice concerning ship routing in Baffin Bay and Davis Strait and the effects of such traffic on the marine and socio-economic environment.

From 1979 until the last working group meeting in 1984 (almost two years after NEB deferral of the project) the review process grew, both in cost and political complications. The environmental studies requested by Danish representatives and paid for by the APP ranged from literature searches of seal and whale data to walrus-tagging field studies and (extremely inefficient) aerial surveys of marine mammal distributions in Baffin Bay and Davis Strait, costing more than $350,000 over two years.

Even more frustrating, however, was the evolution from supposedly scientific evaluation to political badgering. The APP, as a result of questionable environmental concerns, suddenly became a platform for such political missiles as Greenland Home Rule, offshore jurisdiction and sovereignty, native land claims, and the Inuit Circumpolar Conference's mandate to develop a united Inuit voice from Alaska through to Europe.

One need consider only that the Greenland Home Rule Parliament, in the fall of 1981, voted overwhelmingly to oppose the APP and to carry its concern to the United Nations Law of the Sea hearings in New York. The tone of the debate leading up to this vote is illustrated by the following statements: "First, I want to emphasize that the APP is the first project to demonstrate in no uncertain manner, that Greenland and the Greenland society is threatened and to a very high degree. No other project has threatened our hunting and fishing industries, as well as our production sector dependent on them, to an extent comparable to the APP" (Joelsen 1981). "We have seen sufficient instances of the disastrous consequences of establishing reservations for the displacing of ethnic groups. One serious consequence of the implementation of the APP will in all probability, be that the hunting people will have to leave the small communities in order to go hunting in special reservations, and we do not want this to happen" (Andreassen 1981).

These statements were delivered in 1981, prior to any study results being available from the APP working group. They were made on behalf of a country used to offshore hydrocarbon exploration. The statements refer to a project having but two ships, passing through an area approximately every one and a half to two weeks, and located at least 100 km from Greenland's shores.

The complexities of the issue were well stated by Franklyn Griffiths: "Petro-Canada's Arctic Pilot Project presents the intersecting issues in concentrated form. They involve the application of advanced technology, inter-state diplomacy between Canada and Denmark, transnational Inuit cooperation, technology impact assessment, and environmental protection" (Griffiths 1979).

If the APP and the oil and gas industry in general weren't already cynical concerning the motives and sincerity of the Greenlanders, such feelings certainly surfaced with the announcement in September 1984 that Greenland was about to embark on a five-year, $100-million (US) oil exploration program on its own eastern shores (*Calgary Herald*, 27 September 1984). Exploratory work was to be conducted by the American-based Atlantic Richfield Co. of Los Angeles, and oil, if found in quantities sufficient to warrant development, would be transported to market by tanker. Peter Burnet, spokesman for the Canadian Arctic Resources Committee, observed: "It seems awfully strange that they [Greenland] would oppose development in the Canadian Arctic, but then turn around and embrace it in their own country" (*Calgary Herald*, 27 September 1984). D. Bruchet, former assistant project manager for the APP, commented sadly: "We bent over backwards to accommodate Greenland concerns about the environment."

The National Energy Board (1980–2)

Prior to commencement of the National Energy Board's (NEB) formal hearings into the APP in the spring of 1982, environmental and socio-economic aspects of the project had been reviewed by a number of groups. In addition, the NEB was aware that the Coast Guard's Shipping Control Authority and its environmental advisory committee were also about to begin formal review of the APP.

Thus, by the end of 1981, thousands and thousands of pages of scientific documentation and expert testimony relating to environmental and socio-economic aspects of the project had been reviewed by numerous agencies of the federal, provincial, and territorial governments. It was not, therefore, expected that the NEB would start from scratch, yet this is exactly what happened. The NEB, with its history and experience in royalties, taxes, and pipeline technicalities, decreed a full inquiry into *all* aspects of the APP, including social and environmental issues. The process that was to follow would cause the APP's proponents to look back at previous reviews with fondness.

In January 1979, the APP filed an application under section VI of the National Energy Board Act for a licence to export gas from Canada. This application was subsequently withdrawn, and a revised one was submitted in October 1980. The new application consisted of seven volumes, with volumes 5 and 6 devoted exclusively to environmental, socio-economic, and Canada benefit issues. Volume 5, *Public Interest*, was in excess of 700 pages and included socio-economic, environmental, and safety information. Extensive supporting documentation consisted of environmental and social studies undertaken by the APP following the federal EARP report of October 1980, including major field and literature studies on such issues as underwater noise, ringed seal pop-

ulations and distribution, ship-track crossings, Strait of Belle Isle seal
hunt, hunting statistics for caribou, whales, and polar bears, and aerial
surveys of marine mammals in Baffin Bay. The APP had even continued
attempts to develop a more efficient harpoon that would reduce harvest
losses due to sinking. All such information was used to update and
improve the environmental documentation that had been submitted to
the EARP. In total, seventy-two major environmental studies, surveys
and reports were submitted as exhibits to the NEB between October
1980 and the hearing adjournment, in August 1982. A list of document
titles and exhibit numbers is attached as Appendix B.

The APP was also required to respond to eleven "Request for Information Letters" and several more "Deficiency Letters" from the NEB.
These were not simple one-item information requests, but lengthy
packages of environmental queries. The APP had to assign one person,
full time, to catalogue requests and submit staff and consultant responses. In addition, TransCanada Pipelines, responsible for the southern receiving terminal, had to apply for permission to construct the
receiving facilities. All this activity took place prior to outset of the
formal hearings.

In February 1982, almost three years after the original NEB application
was filed, the board's hearing into the APP began. The hearings took
place, in two phases, from February to mid-August. Phase I involved
gas markets, supplies, and economics; phase II, environmental and socio-
economic matters. Only the latter phase is discussed here.

The hearing process involved separate industry panels for separate
issues. For instance, during phase II, separate panels were established,
on Bridport Inlet, ship safety, Canada benefits, and a combined topic
of underwater noise/public interest/special interest. In each case, the
APP was required to assemble a panel of experts to discuss and answer
questions concerning the topic at hand. These experts would individually submit sworn testimony concerning their background and scientific or policy positions. The entire panel and its testimony would
then be subject to cross-examination by board counsel and by the legal
counsel of project intervenors. Both the proponent and intervenors were
required to submit testimony and be available for cross-examination.

Of all the reviews and inquiries into the social and environmental
aspects of the APP, the NEB's was by far the most repetitive, uncontrolled,
and inefficient. The board required that all evidence, cross-examination,
and information exchange must be conducted by legal counsel. Such a
system is better for examination of export taxes, royalties, and pipeline-
compression facilities than for environmental or socio-economic issues.
The latter issues are inexact and cannot be treated the same as a question
of mathematics or pipeline design. A strict judicial review of such issues

as Inuit unemployment or whale migrations curtails the opportunity for meaningful dialogue.

Many times during the hearings, it was obvious that information from both intervenors and the proponent was being poorly delivered by counsel. When scientific advisers have difficulties following what is being said, there is little hope for an informed decision from board members who, themselves, have no background in environmental science.

The intervenor's legal counsel continously attempted to prey on the emotions and sympathies of board members. Inuk after Inuk from the Canadian Arctic, Labrador, and Greenland was paraded before the board, and all pleaded for a return to traditional ways. Science was replaced with tears on several occasions. Industry's position was made impossible. To cross-examine or to bring attention to discrepancies in testimony would only be seen as harassment of witnesses. In most cases, therefore, industry kept cross-examination brief and non-controversial. Reconfirmation of residences and hunting status was typical of the line of questioning. Such a ploy, although necessitated, was not very satisfying to either side or to the search for knowledge.

In August 1982, the NEB adjourned its inquiry, because of changes in the project's export market not because of environmental factors. Since that decision, the APP has been indefinitely shelved by proponents. It is unfortunate that the NEB never rendered a final decision. It would have been interesting to see how such a formal and technically oriented body would have ruled on the environmental and social portions of the project and, especially, to compare its decision with those of such environmentally mandated review groups as EARP and TERMPOL.

POSTURES AND EMOTIONS

A substantial chasm exists between the attitudes of the oil and gas industry and the northern regulatory review process. This void, caused primarily by an unworkable bureaucratic system, not only affects the representatives of the oil industry but also severely restricts the northern native, the non-native, and the government official charged with enforcing the regulatory system. This section highlights industry's growing concerns regarding present and future relations with government and the people of northern Canada. It also touches on industry's understanding, however superficial, of opponents' desires and frustrations.

Historically, the oil and gas industry and its regulators stood on opposite sides of the environmental fence. The situation has changed since the mid-1970s. By building on the early Mackenzie Highway and Pipeline experiences and the surge of Beaufort Sea initiatives, the oil and gas industry realized two things. First, the arctic environment, with its permafrost, sea-ice, and native way of life, was vastly different from the foothills of Alberta or the oil-rich deserts of Texas. Second, but just as important, industry realized that it had a lot to learn about doing business with governments and a country that placed a priority on protecting the northern environment. For these two reasons, the late 1970s and early 1980s saw an important merging of previously divergent approaches. Industry raided government departments in order to acquire instant environmental and socio-economic expertise, and the balance of environmental knowledge, and initiatives in northern study programs, shifted from Ottawa to Calgary.

Although the boom years have passed, and many of industry's instant environmental experts have returned to government or to private consulting organizations, industry retains much improved understanding and a corps of knowledgeable environmental scientists and administrators. Many such individuals will be even more valuable to future industry-regulatory interactions, because of their growing awareness of how both sides operate.

Industry still feels, however, significant distrust, confusion, and frustration toward government and the environmental review regime. The regulatory review process emphasizes negative aspects of development and does little to encourage participants to deal with potentially positive aspects of a development proposal. Both government and the scientific community demanded a small-scale shipping program to safely open the north to year-round shipping. When the APP responded with as safe a shipping project as could possibly be envisaged, the government buried the project with sweeping demands for massive, long-term Canadian content and employment, thereby destroying any chance of a viable pilot-scale undertaking.

Industry lacks confidence in the regulatory system and its managers. Examples abound of ill-founded statements that might never have been made if reviewers were held more accountable for their actions. One recurring concern during the APP review was that the shipping corridor through Lancaster Sound and the Strait of Belle Isle appeared to be very narrow. In most cases this "appearance" was taken from viewing 1:500,000-scale wall maps. It was difficult to convince sceptics that the corridors were many kilometres wide and did not present any unique navigational difficulties.

In making its viewpoint heard, industry often finds itself between a rock and hard place. Especially regarding social and economic effects, the large corporation is often considered as having so much economic and legal leverage as to prejudice reviewers automatically.

Inconsistency of regulatory standards is another source of frustration. We can compare exploration activities in the Beaufort Sea and Lancaster

Sound. The Beaufort Sea EARP review was initiated after almost twenty years, when 150 exploratory wells had been drilled in the Beaufort Sea–Mackenzie Delta region (FEARO 1984). In contrast, the Lancaster Sound Exploratory Drilling EARP was required before initiation of the first exploratory well. Panarctic Oil has been drilling since 1967 in the High Arctic and has spent more than $350 million of its own money or $750 million of joint-venture monies in discovering 18 trillion cubic feet of gas and 500 million barrels of oil ("Frontier" 1984). All this activity by Panarctic has been conducted without formal environmental review.

The greatest affront to systematic project review, however, must be the government's treatment of the attempt in 1983 by the CMO Consortium – Consolidex Gas and Oil, Magnorth Petroleum, and Oakwood Petroleum – to reinitiate exploratory drilling in Lancaster Sound. The CMO Consortium, hoping to find a potential "elephant" of an oil reservoir, applied in 1983 to the Canada Oil and Gas Lands Administration (COGLA) for a deep-water well in Lancaster Sound, planning to start drilling in 1986 or 1987. By 1983, CMO's participants had spent $14 million on seismic research, environmental studies, and regulatory reviews ("Drilling" 1984).

Much of this work had taken place in the late 1970s, before a proposal by Norlands Petroleum to drill a single exploratory well was reviewed by a federal environmental assessment panel. Although the Norlands work was far from perfect, the panel in its final report (1979) decided that, given the many outstanding social and "best-use" questions, "for it to make a recommendation in favour (or not in favour) of exploratory drilling at this time would be arbitrary" (FEARO 1979). The panel therefore recommended a major public review of the best-uses of the Lancaster Sound region, an exercise that took five years (1979–84).

In 1979 as well, industry and government initiated the Eastern Arctic Environmental Studies Program (EAMES). This program, which lasted several years and produced twenty-seven data reports and six integrated analysis reports, was funded by industry. The CMO project picked up a substantial portion of the costs and helped manage the study.

All these activities were undertaken to increase physical and social knowledge of the Lancaster Sound region. Industry continued to participate in the 1980s, believing that such information and patience would be worthwhile in the long run. In 1982 and 1983, CMO undertook to update and improve documentation on environmental impact prepared in the mid-1970s. The *CMO Resource Management Plan* involved a large undertaking of funds, many environmental and social consultants, and many public meetings and consultations. It was submitted to COGLA in the fall of 1983 as part of CMO's request to resume exploration.

In July 1986, the minister of DIAND concluded an exploration agreement with CMO. Thus proponents of exploration in Lancaster Sound waited years and spent enormous sums of money solely to obtain rights to exercise their original exploration permit options.

The Lancaster Sound events point out another difficulty in industry-government relations. Industry is often so unfamiliar with the environmental bureaucracy that it unwittingly becomes manipulated. By being passive and dealing with environmental questions at a non-policy level, industry tends to be swept away by myopic scientists and researchers.

If senior managers approached deputy ministers and ministers, they might define terms of reference more accurately. Representation of program factors by industry's senior personnel might also minimize misinterpretation, which is exacerbated by the time required to move a proposal through the multiple levels of bureaucracy before it ever reaches ministers and deputy ministers. More affirmative senior control of the review process, for example, could have averted an embarrassing meeting of the Regional Screening and Coordinating Committee (RSCC) of EARP looking into the revised Lancaster Sound Exploration proposal. The author of this chapter was present in Calgary when the RSCC met senior management of the proposal. It was learned that, out of about a dozen regulators from federal departments at the meeting, at most two had read any part of the extensive documentation provided. The individual expressing the most concerns represented the forestry division of Environment Canada! Forestry had, until this time, never been a concern in the arctic tundra of Lancaster Sound.

Responsibility should be a primary concern of both industry and government. In a presentation of March 1983 to a group of northern planners, the director general of DIAND's Northern Affairs Program used the following quotation from the February 1983 issue of *National Geographic* as the basis for requiring full native participation in land use planning: "When and if they gain control of their land, the Canadian Inuit will furnish the missing link of a broad resource rich semi-nation of Inuit stretching from Alaska to the North Atlantic. Such an alliance would not only affect the future of petroleum exploration in the Arctic but could also influence western military security and prove a significant fourth world voice in the United Nations" (DIAND 1983a). With heady philosophies such as this, it becomes difficult to convince a drilling manager that there is a practical and responsible side to northern regulatory policies. The same director general stated: "If our society is to avoid suffocation, or the destiny of dinosaurs, it would be wise to return some power to local communities. It would certainly be wise also to preserve large tracts of land, even entire regions, from the enemies of conservation" (DIAND 1983b). Hardly a basis on which to build trust and co-operation!

The oil and gas industry is singled out for special treatment as regard marine transport. Of all the ships that currently travel arctic waters–research vessels, bulk-oil resupply tankers, ice-class bulk freighters, Coast Guard ice-breakers, and military submarines – none has ever had to submit to environmental review. Satisfying the Arctic Waters Pollution Prevention Act Regulations is the limit to which these vessels must conform. Many of the issues and environmental threats associated with commercial oil and gas transport could also be raised for other shipping. Such concerns as underwater noise, disturbance of bird and marine mammal populations, alteration of natural ice regimes, and the possibility of vessel fuel spills could be pinned to all vessels operating in the north. It would be interesting to see, for the sake of equity, how aging Coast Guard ice-breakers or even Inuit outboard motors would fare in a full-blown environmental review.

RECOMMENDATIONS

Many authors have criticized the northern environmental regulatory process. Very few have suggested practical methods for beneficial change. The system is perhaps too firmly rooted, in time and bureaucracy, to allow piecemeal alterations. The alternative, therefore, is a major operation that examines the entire system of northern environmental review. Such an exercise could minimize duplication and contradiction and maximize co-ordination and efficiency of review.

To have any chance at success, such a review would have to be conducted at a very high level, with well-defined terms of reference and deadlines. All departments of government having an environmental mandate and many departments having economic-development and social mandates would have to be involved. Northern industry interests, from oil and gas to small business, would play major review roles. Members of the general public and responsible special-interest groups would also participate. Legislation and regulations would have to be changed, and politicians would have to be willing to sacrifice personal and departmental profiles and powers.

Being unable to envisage the perfect system, this author, like many before him, can offer only piecemeal suggestions on how to streamline the overall system. The first suggestion is simply a call for increased awareness and understanding of the industrial development process. Most mega-scale projects change and evolve from initial concept to final design. The option must always exist to incorporate changing engineering, economic, and scientific parameters into the project description. Lengthy public reviews only increase the probability of design changes. If regulatory bodies could accept such a philosophy, industry would appear far less deceitful and contradictory in the submission of project information. It would also become clear why companies are loathe to accept final-design parameters early during a review. Final design is extremely complicated and expensive. Private industry cannot undertake this step without some assurance that the project will proceed.

Inter-industry competition and confidentiality are also not given proper recognition and understanding. In many review forums, industry is accused of withholding information. In many cases, the accusation is valid and readily admitted to on the basis of confidentiality. If the engineering design of a ship-propulsion system were unique but as yet unpatented, it would be unbusiness-like to reveal design details in a public hearing. Company's projections of profit and plans for future projects are regularly refused to reviewers, resulting in public resentment. The backbone of business is the ability to get ahead of the competition, and industry should not be penalized or maligned for valuing proprietary information.

In conclusion, government and industry can and should work together toward a better-organized and more concentrated process of environmental impact assessment. This need has been identified in a recent study by government, university, and industry: "A forum for productive discussion and the exchange of ideas among those administering, conducting, reviewing and paying for impact assessment studies must be established. Resolution of the principal difficulties will be slow unless the major participants are aware of more than just the problems inherent in their own responsibilities" (Beanlands and Duinker 1983).

In other words, a joint industry-government forum is needed to work co-operatively, throughout the assessment process, beginning in the project-development stage and continuing through to final project definition. Such a group would be much better equipped to identify priority information or design requirements and the means to satisfy them. In addition, by combining talents, members would be able to discard irrational or unjustified concerns and thus take a giant step away from the traditional grab-bag collection of baseline data. As they focus more on prediction, mitigation, and monitoring of impact, study boundaries will automatically identify relevant gaps in baseline data. A joint working group will bring issues into an arena where they can be legitimately and rationally assessed and thus might even result in respect between two traditional opponents.

APPENDIX A

Vessels Crossing 60°N to and from the Arctic in 1980

Month	Type	Vessel size (grt)			
		500– 2,999	3,000– 9,999	10,000– 19,999	20,000 and up
TO THE ARCTIC					
July	Tankers	2	3	1	–
	Others	15	6	1	1
August	Tankers	1	6	3	1
	Others	13	6	2	4
September	Tankers	–	–	1	–
	Others	5	2	4	3
October	Tankers	–	–	–	–
	Others	5	1	2	2
Total		41	24	14	11
FROM THE ARCTIC					
August	Tankers	3	3	3	–
	Others	12	3	4	3
September	Tankers	–	6	6	–
	Others	12	5	3	5
October	Tankers	–	1	–	–
	Others	7	4	2	3
November	Tankers	–	–	–	–
	Others	4	2	–	–
Total		38	24	18	11
Grand total		79	48	32	22

From *Annual Report Coast Guard Ice Operations – Arctic 1980.*

APPENDIX B

APP Environmental Documents Filed with the NEB

Exhibit No.	Description	Date
32	Volume 5 – Public Interest (As Amended)	82-02-02
35	Environmental Atlas	82-02-02
	ACRES CONSULTING SERVICES LIMITED	
51	Environmental Impact of Thermal Discharge. July 1979	82-02-02
52	Ice Management within Bridport Inlet. Executive Summary Report. July 1978	82-02-02
54	Studies on Ice Management – A Summary Report. December 1979	82-02-02
	ARCTEC CANADA LIMITED	
75	Study of Bridging a Ship Track in Landfast Ice. April 1979	82-02-02
76	Study of Influence of Shipping on Break-up and Freeze-up in Lancaster Sound. December 1979	82-02-02
	ARCTIC PILOT PROJECT	
81	Environmental Overview. Gas Production Component	82-02-20
82	Environmental Statement. Melville Island Component	82-02-02
83	Environmental Statement. Shipping Component	82-02-02
84	Environmental Statement. Supplementary Information. November 1979	82-02-02
87	Relative Merits of Vessel Transit through the Strait of Belle Isle or around the East Coast of Newfoundland to Melford Point, Nova Scotia. August 1981	82-02-02
92	Summary Environmental Statement	82-02-02
	ESSA (ENVIRONMENTAL AND SOCIAL SYSTEMS ANALYSTS LIMITED)	
104	Towards an Environmental Management Strategy for the Arctic Pilot Project. March 1980	82-02-02

Exhibit No.	Description	Date

FROZEN SEA RESEARCH GROUP
An Oceanographic Study of the Bridport Inlet,
Melville Island, N.W.T.

109	Part I – August 1978	82-02-02
110	Part II – March 1979	82-02-02
111	Part III – March to August 1979	82-02-02

HAMILTON, E.G. AND BLISS, L.C.

117	Active Layer Detachment Slides on King Christian Island and the Sabine Lowland in the High Arctic. February 1979	82-02-02

R.M. HARDY & ASSOCIATES LIMITED

118	Landscape Survey – Eastern Melville Island, N.W.T. April 1978	82-02-02

HATFIELD CONSULTANTS LIMITED

119	Environmental Assessment of Selected Freshwater Resources near the Proposed Petro-Canada LNG Project, Melville Island, N.W.T. November 1977	82-02-02
120	Observations of Marine Mammal and Seabird Interaction with Icebreaking Activities in the High Arctic, July 18 to August 5, 1979. December 1979	82-02-02
121	Observations of Marine Mammal and Seabird Interaction with Icebreaking Activities in the High Arctic, July 18 to August 5, 1979. September 1980	82-02-02
122	Observations of Marine Mammal and Seabird Interaction with Icebreaking Activities in the High Arctic, July 2–12. December 1980	82-02-02

INTERA-ENVIRONMENTAL CONSULTANTS LIMITED

125	Remote Sensing Overflight. Data Acquisition and Reduction Program. November 1977–May 1978	82-02-02

LGL LIMITED

126	Late Winter Distribution of Black Guillemots, *Ceppus grylle* in Northern Baffin Bay and the Canadian High Arctic. January 1980	82-02-02
127	Marine Mammals Inhabiting the Baffin Bay North Water in Winter. January 1980	82-02-02
128	Numbers and Distribution of Birds on Eastern Melville Island, July to August, 1977. December 1977	82-02-02

Exhibit No.	Description	Date
129	Numbers and Distribution of Marine Mammals along the Coast of Southeastern Melville Island, July to August 1977. November 1977	82-02-02
130	Studies of Terrestrial Mammals on Eastern Melville Island, July to August, 1977. December 1977	82-02-02
131	Survey of the Marine Environment of Bridport Inlet Melville Island. December 1977	82-02-02
	NORDCO ENGINEERING AND RESEARCH LIMITED	
148	Proposal for Winter Baseline Studies at Bridport Inlet. September 1977	82-02-02
	SCHLEDERMANN, PETER	
160	1979 Archaeological Site Survey of Bridport Inlet and the Proposed Interior Pipeline Corridor, Melville Island, N.W.T. October 15, 1979, and Addendum, February 1980	82-02-02
	TOP, ZAFER	
161	Helium-Tritium Analysis, Bridport Inlet, Melville Island, N.W.T. November 1978	82-02-02
	WESTERN RESEARCH AND DEVELOPMENT	
162	The Air Environment of the Proposed Natural Gas Pipeline Route on Melville Island, N.W.T. December 1977	82-02-02
163	An Analysis of the Air Environment of Bridport Inlet, N.W.T. January 1978	82-02-02
164	Comparisons of Wind and Temperature Data Collected on Melville Island at Bridport Inlet, Beverley and Rea Point over the Period July 15, 1977–May 31, 1978. September 1978	82-02-02
167	Integrated Route Analysis – Volume I	82-02-02
168	Integrated Route Analysis – Volume II	82-02-02
169	Integrated Route Analysis – Volume III	82-02-02
170	Transport Canada Termpol Assessment of Melford Point – Volume I	82-02-02
171	Transport Canada Termpol Assessment of Melford Point – Volume II	82-02-02
172	Transport Canada Termpol Assessment of the Arctic Pilot Project (Northern Component). August 1981	82-02-02

Exhibit No.	Description	Date
173	Transport Canada Termpol Assessment of the Arctic Pilot Project Southern Component (Gros Cacouna Alternative). September 1981	82-02-02
174	Transport Canada Composite Executive Summary to Independent "Termpol Code" Assessment of the Three Arctic Pilot Project Sites Submitted	82-02-02
175	The Question of Sound from Icebreaker Operations Proceedings of a Workshop 23, 24 February 1982, sponsored by APP	82-02-02
176	Report of the Environmental Assessment Panel – Arctic Pilot Project (Northern Component)	82-02-02
177	Report of the Environmental Assessment Panel – LNG Receiving Terminal (Arctic Pilot Project) Melford Point, Nova Scotia. August 1981	82-02-02
178	Bureau d'audiences publiques sur l'environnement/ Report on an Inquiry and Public Hearings – Gros Cacouna LNG Terminal Translated by: FEARO	82-03-11
199	Acres Consulting Services Ltd: Ice Crack Morphology in Barrow Strait. July 1981	82-02-02
200	LGL Limited. Distribution of Wintering Marine Mammals in Southern Baffin Bay and Northern Davis Strait. March 1981	82-02-02
201	Arctic Pilot Project's Comments Regarding the Termpol Coordinating Committee's Independent Executive Summary. October 1981	82-02-02
202	Arctic Pilot Project's Comments Regarding Nova Scotia EARP Report	82-02-02
203	Arctic Pilot Project's Comments Regarding Quebec EARP Report	82-02-02
204	T.G. Smith and M.O. Hammill: Ringed Seal, Phoca hispada, Breeding Habitat Survey of Bridport Inlet and Adjacent Coastal Sea Ice. July 1980	82-02-02
322	A Study of Sealing Activity in the Strait of Belle Isle and off Labrador. Nordco Limited	82-02-04
349	Transcript of Proceedings of Federal Environmental Assessment Review Panel in the Matter of the Arctic Pilot Project. April 23, 1981	82-02-05

Exhibit No.	Description	Date
397	Document Entitled: "Proposed Organization of a Research and Development Program for the Arctic Pilot Project – February 10, 1982," Prepared by Pallister Resource Management Ltd., Calgary Alberta	82-02-24
507	Document Entitled, "APP Documents Which Were Passed to the Chairman of the TERMPOL Environmental and Socio-Economic Sub-Committee (Northern Component)"	82-04-19
531	APP Document Entitled, "Preliminary Oil and Hazardous Chemical Spill Prevention and Contingency Plan," Prepared by the Environmental Protection Group, Petro-Canada	82-05-10
532	Study Prepared by LGL Limited Entitled, "A Study of Muskox Behaviour and Distribution during Early Rut on Eastern Melville Island, August 1981"	82-05-10
544	Report Entitled, "Icebergs in the Strait of Belle Isle and Approaches"	82-05-11
548	Report of Netherlands Ship Model Basin, Dated February 1982 Entitled, "Radiated Noise Tests for an Arctic LNG Carrier"	82-05-12
557	Chart III-D-5 Figure 1, Entitled, "Estimated Trip Voyage Duration for Two Vessels"	82-05-13
570	Document Headed, "Ice Effect Trials in Arctic Waters on CCGS *Louis S. St. Laurent*"	82-05-19
616	Chart of the Strait of Belle Isle Bearing No. 4020	82-06-07
622	Revisions to the Integrated Route Analysis	82-06-08
668	APP Document Entitled, "A Regional Analysis of Fishing Activity from the Labrador Sea through the Strait of Belle Isle to Cape Ray," in Response to an Undertaking Given to Mr. Haysom at Page 9594 of the Transcript	82-07-05
672	Document Entitled, "Replace Table 1 (p. 65) and Table II (p. 68) of Prepared Testimonies, Phase II. Panel 6A – Noise, with the Following Revised Tables: Table I. Measured and Computed TL for Baffin Bay at 100Hz; Table II. Received Noise Levels from an APP Carrier in Baffin Bay in Open Water at 100Hz as a Function of Range"	82-07-05

Exhibit No.	Description	Date
673	APP – Material Entitled, "Sources of Underwater Ship Sounds: Flow Noise, Machinery, Turbines, Gears, Pumps, Propellor Cavitation, Icebreaking. Propellor Cavitation, the Dominant Source of Ship Sounds Underwater"	82-07-05
675	APP Document Entitled, "Table II: Received Noise Levels from an APP Carrier in Baffin Bay in Open Water at 100Hz as a Function of Range"	82-07-05
685	APP Document Entitled, "Ship Track Crossing"	82-07-07
714	Document Entitled, "Observations of the Bowhead Whale (*Balaena mysticetus*) in Central West Greenland in March–May, 1982," Prepared by E.W. Born and M.P. Heide-Jorgensen	82-07-12

CHAPTER SIX

Inuit Concerns and Environmental Assessment

Peter Jull

In August 1985, a remarkable situation unfolded in the Canadian media. An American ice-breaker sailed from Thule in Greenland, through the Northwest Passage, to the Beaufort Sea. The ship was armed and rumoured to be engaged in military research. This voyage aroused Canadian fears about national sovereignty and jurisdiction in the Arctic, fears only enhanced by the evident inability of Canadian cabinet ministers and officials to explain the government's position.

Inuit spoke out concerning the threat this voyage posed to the marine environment and the sea ice which are central to their food and other requirements. They said that they supported maximum Canadian control in the area and wanted to be assured that Canada was able to protect their interests. At a panel discussion in Ottawa at the time, one panellist called for greatly increased Canadian-organized shipping in the Arctic to demonstrate sovereignty. The Inuit international spokesman, Mark R. Gordon, responded that Inuit hunters and families living and travelling on the sea ice and hunting the sea species through the centuries provided Canada's best evidence of sovereignty. "We will keep the flag flying for you all!," he concluded, to a round of applause.

Inuit had refused a hastily offered concession by the federal government that an Inuk observer be placed aboard the ship as it passed through Canadian waters. The board of directors of the national Inuit organization – the Inuit Tapirisat of Canada (ITC) – considered that such an action would be meaningless and would appear to indicate Inuit acceptance of the ship's passage. Later the government dispatched one of its own Inuit employees to the ship.

This episode reveals the main problems of environmental management in the Inuit homeland of arctic Canada. Canada recognizes that Inuit have legitimate interests and cannot be ignored, but ad hoc responses in moments of political embarrassment are geared more to the appearance of sensitivity to Inuit needs than to practical purposes. Inuit defend their basic economic interests in the region, but modestly and with sincere regard for national sensitivities. Nevertheless, the actual powerlessness of the Inuit is all too evident. And the ice-breaker sailed on as the federal government appointed a study group to inquire into Canadian sovereignty!

Two other points about this incident need to be noted. First, the ship had first visited Thule, from which Inuit were removed in the post-war era with great loss of traditional and productive territories (Brosted and Faegteborg 1985). The ship was bound for Barrow, Alaska, where the international organization of Inuit – the Inuit Circumpolar Conference (ICC) – was set up in 1977, to deal with such threats as this to the marine environment of the Arctic. Second, the episode was a repeat of the voyage of the *Manhattan* sixteen years earlier. Yet this second American challenge was known well in advance by Canadian officials, and Canada's response could likewise have been thoroughly prepared in advance. One can conclude from all this that Americans, likely the main users of the seas of arctic Canada in the future, have little regard for Inuit sentiments and also that the Canadian government is unable or unwilling to protect the Inuit homeland and the marine environment.

The 1985 case may be exceptional in the level of public interest it aroused, but it is part of a pattern. Earlier in 1985, for example, the federal government, at a press conference in Yellowknife, announced approval of the Bent Horn oil project. Oil from this project would be transported through Lancaster Sound. Tankers would then move through Davis Strait, where the Greenland fishery, the main economic base of that country, is located.

Inuit communities and associations had generated great publicity and put pressure on federal officials a year earlier to demand protection of Inuit interests in both Canada and Greenland in relation to Bent Horn. Now, the official press release made only the vaguest promise of consultation with Inuit on industry plans. Two ministers from the Northwest Territories (NWT) government, both of them Inuit, were involved in the press release, and one was noticeably concerned about the effects on marine life. The more enthusiastic minister of economic development, well known and controversial among Inuit for his support of large-scale resources development, gained more attention, however. The press release appeared the evening before the federal government announced approval in principle for creation of an Inuit government in Nunavut, the Inuit homeland in eastern and central NWT, in a move timed to lessen criticism of the oil project in the euphoria of political reforms achieved.

One wonders about the competence, not to mention the sincerity, of those in Ottawa who make decisions for the north. Often they are the first to claim to be acting in "the national interest" – but do they know what that interest is?

BACKGROUND

The ancestors of the present Inuit migrated from the Bering Sea into and across the Canadian Arctic in the period AD 800–1000. Primarily a whale-hunting culture, they were merely the most recent immigrants from a region that had supplied wave after wave of settlers to the Greenland and Canadian arctic regions following the last Ice Age. As the climate deteriorated and colder years set in, the whales moved elsewhere and Inuit had to adapt to other local resources and conditions. In the Inupiat Inuit homeland of northern Alaska, whaling is still the centre of life, though even that vigorous and dynamic society is taking on many present-day challenges.

Today's Inuit are no less marine-oriented than those in the past (see Cooke and Van Alstine 1984). In Greenland a sophisticated fishery has replaced the traditional life-style, but much of the old outport life persists. In Greenland and Alaska, and apparently the Soviet Union as well, harvesting of traditional species by Inuit continues.

But Inuit in Canada hold little power in relation to the marine environment or its protection. Jurisdiction over offshore matters is federal and has never been lightly shared with maritime regions and their governments. Inuit do have some influence in the NWT government, but ocean interests have not been a priority with it. Indeed, during the referendum of 1982, in which Inuit sought (successfully) a majority vote for dividing NWT to create Nunavut, a major argument advanced was that Inuit were a maritime people who needed a more maritime-oriented government to promote and protect their interests. Yet Inuit have been denied higher education and have lacked facility in Canada's official languages and access to the news media and to federal Ottawa and to government officials. They have had few opportunities to promote their view.

Since the early 1970s, however, the federal government has provided funding to all aboriginal peoples across Canada to help them present their concerns to public authorities. This level of support is considerable and evokes admiration in other countries. In the case of Inuit, it has meant creation of a national organization, Inuit Tapirisat of Canada, with headquarters in Ottawa, and regional organizations in each of the six "Inuit regions" – Labrador, northern Quebec, Baffin, Keewatin, Kitikmeot (central Arctic), and the "COPE region" (Mackenzie River Delta and Beaufort Sea communities). More recently, Inuit television broadcasting services and a constitutional affairs group (the Inuit Committee on National Issues, or ICNI) have also been able to deal with specific issues.

Inuit have expected these regional and national associations to contend with threats of environmental degradation as a result of oil and gas exploration. These bodies have thus spent a great deal of time and money on such issues. Their record has been uneven, but they have made the public and governments aware of Inuit concerns. They have also developed various viewpoints and amassed data that have been as useful to the nation as a whole as to the Inuit – for example, arguments in international law for demonstrated "Canadian" use of arctic lands and various scientific lines of inquiry in relation to behaviour of sea mammals. In co-operation with Greenland and Alaska Inuit, they have created an international body, the Inuit Circumpolar Conference (ICC), and held four well-publicized week-long assemblies, at which environmental and ocean issues were discussed.

Yet even though government has had access to Inuit views – through hearings, community visits and consultations, special-purpose consultative bodies, and so on – these have had only limited effect. True, major development has not yet begun in the north, despite many threats. But Inuit protests are only partly responsible for this situation, world markets and technological deficiencies having been more of a contributing factor.

The time-consuming hearings and review that the government has set up to address concerns have not followed a consistent pattern. They have taken many forms, and each has had a personality of its own. There has been no predictable outcome, and many dollars have been spent collecting and preparing research and opinion for presentation, with the same issues being covered again and again. Inuit have not felt relaxed with this process, nor do they ever know that their case has been effectively or finally made.

Before discussing the broad issues, it may be useful to look at the contents of one such inquiry, as a typical example of the process and to obtain the flavour of the issues.

A CASE STUDY: THE BEAUFORT EARP

The most recent, most expensive, and most comprehensive study, dealing with the whole width of the Arctic, was carried out by the Environmental Assessment Review Process (EARP) Panel established in 1981 to look at production and various modes of transport, including ocean tankers, of Beaufort Sea hydrocarbons. Both the Beaufort panel and process, and later its report (July 1984), came to be known by the undignified acronym BEARP.

At the time of BEARP's arctic hearings, some Inuit leaders wondered about yet another study of issues already exhaustively dealt with in the successful struggle against the Arctic Pilot Project. Yet the shipping of oil through Lancaster Sound at the eastern end of the Northwest Passage was of deepest concern. The sound had been studied many times. But the people of the four nearby communities needed no studies to be made aware of its importance and its role as a major source of their livelihood and well-being. For Canada's environmental community at large also, Lancaster Sound had become a touchstone and a symbol of the government's attitude toward the Arctic in general.

If Inuit of eastern NWT (Nunavut) were sensitive to marine transport, so were Labrador Inuit, along whose dangerous and stormy coasts the tankers would travel. But it was the western Inuit, of the Committee for Original Peoples' Entitlement (COPE) region, who had most at stake. Already suffering considerable social upheaval, including sudden and unfamiliar prosperity resulting from exploration activity in the Beaufort Sea, their entire social, environmental, and political future was at stake. A higher level of activity would, after all, attract enough workers and ancillary activities to overwhelm the small number of Inuit. Inuit ability to influence a future dominated by uncaring transient workers from outside would be slight. Nor were Inuit short of scary stories and grim advice from their neighbours and kin on Alaska's North Slope. But in Alaska Inuit had organized a rough-and-ready government of their own, the North Slope Borough, to benefit from and control the oil industry. In Canada, not only did the oil industry seem to be slowly getting its way, but senior government officials were reluctant to accept any proposals for more than minimal Inuit power in the industry's development.

It was in the Lancaster Sound region and also on Banks Island that the recent era in relations between Inuit and development had begun. Repeated Inuit worries about exploration crews were met with stylized government assurances. When a research report from the Department of Indian Affairs and Northern Development outlined these events and documented the deceptions and misinformation given out by its minister and officials, the conflict began. The author of the report was removed and his superior moved to another position. From that day in 1970 the struggle has continued. It is an enduring and archetypal conflict, which leads not to satisfaction or resolution but to repetition of contrasting philosophies and world-views, to which no one really listens or responds.

As it was, the Beaufort Sea region had already suffered the greatest changes of any Inuit region in Canada. From the days of the New England whalers onward, these Inuit had suffered disease and death, loss of resources and damage to habitat, as well as humiliation, marginalization, and loss of culture. Some people even said that so much had happened that it really didn't matter what more followed. However, it was out of this situation that the first and strong Inuit organization, COPE, was born. It grew out of the social problems in the "model community" of Inuvik, built to symbolize a new and permanent north. The town instead had favoured the educated whites, who held secure institutional jobs, and left the Inuit on the fringes, in unserviced shacks. It became a symbol of discrimination by whites, of social malaise among native Inuit, Dene Indians, and Metis, and of the futility of planning decisions made and implemented by outsiders.

Yet despite social ravages and environmental degradation, as Mr. Justice Thomas Berger found in his landmark report, *Northern Frontier, Northern Homeland: The Report of the Mackenzie Valley Pipeline Inquiry* (1977) (which included intensive studies of the COPE region), the underlying economy of the traditional use of land, seas, rivers, and lakes was still of immeasurably great value to Inuit. That value was not simply economic but was the essence around which social organization and the entire Inuit culture were built. As with other aboriginal peoples, Inuit roots in their homeland and their sense of identity with living resources transcended anything the new settlers or their governments could comprehend. What was remarkable was that a society viewed by white people with either pity or contempt was surviving. Mr. Justice Berger found in fact that the worst was over – a time when the Inuit and other neighbouring peoples had been stunned by the impact and ferocity of change – and that now a renascence was taking place. In this new phase, the northern peoples would be using the white people's own methods and forums and would be absorbing and adapting to many changes. Inuit society would survive, but in reconstituted form.

BEARP's report includes a clear summary of community meetings at which the Inuit position was expressed clearly. Inuit are deeply concerned about social effects, most notably alcohol abuse and disruption of young lives in pursuit of an alien way of life. They worry that people will leave and be alienated from the community; they are concerned especially about direct environmental effects (for example, the effects of ship noise on wildlife) and doubt that developmental proponents have an adequate grasp of these. They want to know about the capacity of industry and government to handle pollution accidents. They recognize, however, that employment opportunities from the project could be beneficial.

Nunavut community meetings brought forward the same worry over employment for young people. Unsuited for traditional life but unable to find work locally or handle the job market "down south," young folk mill around the villages, where video games, television, drugs, alcohol, and family violence provide emotional outlets. How can jobs be provided and the community be kept together and sustain itself both socially and economically, in relation to its traditional environmental base?

As regards social concerns, BEARP is not encouraging, noting that "the linkages between development and social problems are mostly indirect and difficult to separate from other impacts" (FEARO 1984:41). This statement has little meaning, because the only effects in the Arctic are from development by government or industry. The links, furthermore, are not "indirect." They could not be more direct! But coming into the Beaufort and Delta communities in the midst of whirling dislocation and change, one may forgive the panellists their short-sightedness.

BEARP describes the present situation in understated and subjunctive bureacratic style: "Local communities and their residents are, in some cases, ill-equipped to deal with social change and social problems. The additional effects of oil production and transportation activities may compound their problems. The communities will be placed in the even more difficult position of reacting to problems rather than attempting to control them. Social problems and inadequacies in community services may be aggravated by differences in needs among the permanent and the incoming populations" (FEARO 1984:40).

The BEARP chapter on "The Human Environment" generally is reluctant to find causes, to identify those responsible in government, and to believe that problems can be solved, least of all by public action. But in preparing for development (which means "more development"), "part of this preparation will be to help individuals understand what to expect in the future, where government and community responsibility and assistance ends and personal responsibility begins, and how to cope with changes now occurring or which may occur" (FEARO 1984:40). Presumably BEARP's members and many expert advisers, working on the most expensive environmental assessment in history, at $20 million, would not predict or advise on coping with change, because, according to these individuals, that cannot be done. And while eschewing government responsibility for all ills, they would have some insightful person (a public official, of course) move in with a lucky Inuit family and advise on personal responsibility! Comfortably seated in tent, igloo, or bungalow, this official might find his or her expertise on personal responsibility undermined: Inuit men, women, and children distribute responsibility, personal or other, in quite different ways than university-educated southern whites.

As for government and community responsibility, the collective and

consensus decision-making of Inuit has long attracted notice from researchers and government commissions (for example, the Drury Report, on constitutional development in NWT). The political culture of
the Inuit is not the same as that of white southern Canadians, and their
self-government style will probably differ, too.

Yet elsewhere in dealing with the Inuit, this same chapter acknowledges inadequate and insufficient public services in social fields, which
will be exacerbated by further development (FEARO 1984:45). And then
comes a rare insight: "Spending more on social services to help residents
cope with development can only be seen as a supplement to the important task of establishing northern residents as key participants in
development" (FEARO 1984:45). Here, at least, the problems of power
and of participation, two sides of one coin, are identified as contributing
to social ills.

But the insight is not sustained. Soon we are told that governments
should set up careful monitoring processes to watch for inflation caused
by development, "so that government and industry can act swiftly" in
"adjusting the levels of assistance payments" (FEARO 1984:52). And
BEARP is always quick to see the bright side: "There could also be an
enhanced interaction between native northerners and others creating
beneficial cross-cultural social interaction, and opportunities and motivation for travel and higher education" (FEARO 1984:41).

Yet in other respects BEARP is curiously reticent, as in introducing
specific social items: "A number of topics – such as alcoholism, family
problems and crime – are discussed without conclusions or recommendations. These are general problems which require continuing attention" (FEARO 1984:42).

Yes, indeed, these problems are general. They are the story of the
modern north and are especially well documented in a book about oil
and gas development in Alaska among an identical Inuit population,
the North Slope Inupiat. That book, *Eskimo Capitalists: Oil, Politics and
Alcohol*, by Klausner and Foulks (1982), is much more direct and comprehensive when dealing with social issues. Its conclusions, and many
like them, are unmistakable. Indeed, most right-minded adults would
find the evidence of their eyes unmistakable, too, if they were able to
travel around the arctic villages. That is why BEARP's curiously abstract
tone is difficult to comprehend.

The section "Alcohol Abuse" takes up one-quarter of a page – 0.17
per cent of the total text – yet alcohol accounts for perhaps over 60 per
cent of the human effects that will result from projected developments
in the region. "Intervenors from communities indicated that alcohol
abuse is frequent in many northern communities" (FEARO 1984:44).
Apparently the panel took no responsibility for an observation which
they felt obliged to pass on, however, since it came from local people.

On this critical topic BEARP only begs the question: "The Panel believes that, in many cases, alcohol is both the result and the cause of many social problems and concludes that communities must help develop methods to deal with their own problems" (FEARO 1984:44).

While living Inuit are given short shrift, the dead are positively celebrated. In an almost lyrical passage on archaeology, detailed advice is given on the importance, fragility, and care of human remains (FEARO 1984:99). More money is advocated for programs to preserve these, and communities are encouraged to join in the digging. Inuit may someday tell their personal responsibility officer that they think the white person has his or her priorities skewed!

Turning to government structure and operations, BEARP recognizes the lack of Inuit participation: "The Panel was informed that northerners are particularly frustrated by their inability to reach the federal decision makers and with the apparent lack of accountability to the local people of these decision makers. The Panel believes that this frustration may be at the root of many of the socio-economic problems from which northern communities suffer" (FEARO 1984:95).

It recommends more power for territorial governments, though not what powers, a considerable oversight in a constitutionally minded country. And one of its most important recommendations – that "an increased share in resource revenues for northerners would enhance northern benefits and local autonomy, and would serve to make development more acceptable" – is buried in the narrative and not listed among the many numbered recommendations in bold-type (FEARO 1984:97).

In short, BEARP lurches from irresponsibility and negligence to intrusion and condescension in its treatment of Inuit (and Dene and Metis) and betrays petulance and impatience with native peoples. It occasionally acknowledges that northerners must organize their own affairs, largely to facilitate changes that BEARP itself cannot deny (but can neither describe nor prescribe!). It reveals the incomprehension of one society and economic ethic when faced with quite a different society. Its ill-assorted conclusions form a grab-bag of solutions, not a coherent plan, and provide no useful guidance to policy-makers. The overall conclusion, that small-scale development may proceed slowly, is an a priori reaction, revealing confidence in development and hesitation about socio-economic and environmental matters, rather than being a true conclusion based on facts or on the opinions of those likely to be affected.

THE LARGER PICTURE

Whatever the deficiencies of the BEARP report, it would have seemed amazing a few years ago for environmental assessment to be straying into political speculation, constitutional and institutional reform, social philosophy, and ancestor worship. But this "drift" is instructive and reveals what has happened to environmental assessment in the Arctic.

To Inuit, environmental assessment processes, however ad hoc, have become outlets for expression of political aspirations. It is in these forums that Inuit most often state their case. But industrial proposals for arctic development are only symptoms of larger governmental and public attitudes toward development, the use of the natural environment, the north, and the young people of the north.

Inuit have probably been successful in communicating their anxieties and needs. That there is a response to these in southern Canada is evidenced by an investigation of respondents to a direct-mail fund-raising program by the Inuit Tapirisat in 1984. The investigation found only one common motivation: southern identification of Inuit as protectors and promoters of the natural environment's balance and harmony in wildlife and habitat use. This being the way Inuit view themselves, one might call such a shared understanding felicitous. But those who wield power in the north and determine what will happen have not yet accepted this perspective.

It is important to understand what is happening in the north. As far as industry and government are concerned, a project is going into a comfortably Canadian area unhampered by provincial jurisdictions, and an environmental assessment will reassure people and protect whatever sensitive spots may require special treatment. To Inuit, an alien industry is being thrust into their ancient homeland and social milieu – more evidence that their rights and traditions are not respected by the Canadian government. Inuit respond not only to a particular project but with a coherent viewpoint that is part and parcel of a whole way of life. They respond not as a mere interested party, but as a people who have managed their own lives in this, their own territory, since earliest memory. While few if any of them might use the term *popular sovereignty*, it is that sense which they are expressing. Their moderation and lack of demonstrativeness mean that they react quietly to these issues. But they are anxious, and their anxieties are transferred to other areas, notably claims negotiations and demands for self-government and constitutional amendments. These demands include the need to secure political rights, together with protection of their ocean interests (see Jull and Bankes 1984).

That Inuit obtain little satisfaction from their present territorial government is certain. In the hearings around the north to discuss a new Nunavut government, one of three recurring themes was the need for greatly expanded powers in relation to marine matters. The jurisdictional maze of federal Ottawa is too far away, and too strange for easy access. (Nevertheless, more attention to arctic marine issues has been given there in recent years, and now several offices in government are trying to improve their response to Inuit interests, or even use these interests as support for their own more immediate ones – for example, more stringent environmental standards. Also, Ottawa has agreed to negotiate Inuit rights and benefits as part of "land claims.") As with other aboriginal peoples around the world – and the other northern homelands in Canada – politics and the environment are inseparable. Aboriginal politics are environmental politics and represent a fundamental critique of present-day industrial society (see Jull 1985).

Divergent perspectives about land negotiations are apparent. Governments see them as land transactions, and aboriginal peoples view them as fundamental political negotiations covering a broad range of problems. For Inuit, use of the sea and sea ice is as important as land use.

Inuit are an absolute majority of the population across one-third of the land area of Canada. Yet they are few in number – about 30,000 in this country and barely more than 100,000 world-wide. Despite occupation of vast lands, they have little influence in public institutions. Everywhere they are a minority. They are indeed represented in public institutions, which, however, are dominated by others and little understand their language, customs, way of life, and needs. Their influence depends on moral suasion, expert lobbying, and the quality of their research, but they have few opportunities to make decisions. This disadvantageous situation in their ancestral homeland of Nunavut in eastern and northern NWT is the reason for their plans to restructure government and create a new territory in that region (Jull 1986).

Environmental assessment processes take on a very special role in the north. Their formality – and even informal hearings assume a formal guise in the Inuit north – and trappings indicate their importance, and it may be hard for villagers to understand that such costly and solemn proceedings are in fact merely advisory and may result in very little. The white people come to one's own village in order to be accessible. If white people can carry out such vast enterprises to no purpose, their wealth must surely be so great that they do not need the wealth under the arctic seas and lands anyway!

And Inuit are an equal match for any southern experts in these hearings, because they are well informed about effects. Their generation has been moved by government from a traditional life of scattered camps to bungalow communities that might appear to belong more easily in Edmonton than on Baffin Island. This society has never recovered its sense of balance, and with all the impending upheaval – new types of housing, new work or more probably no work, education in a strange language, alien foods, the curse of alcohol, and the introduction of institutions run by young and inexperienced whites who have power over one's life and family – the new communities are not always oases of contentment.

The various southern-style institutions – village councils and other bodies, often with their own British heraldic crests featuring northern lore, such as the "galloping sardines" (narwhals, actually) on the NWT crest – may seem mere pretense, practice for something yet to come. Surely they do not mean to the young white men and women incorporating Inuit culture into these institutions the same things as they do to older Inuit? When more Inuit viewpoints on these matters are recorded, they will be the beginning of the real history of Nunavut. Meanwhile, whatever the limits and drawbacks of locally provided arenas for debate, environment assessment forums are attractive for the expression of deeper feelings, because they deal with the issues that are most meaningful to Inuit.

In environmental forums, too, Inuit can use their old lore, the things they know best, the ways of sea and land, sky and winds, so that those Inuit who are most rustic and ill-suited to the new ways can be respected and listened to by whites. The old Inuit have at least shown that white science is sadly lacking. But they have also revealed the overall integrity of a society which, in more ways than are affected by a ship's passage through ice or by an oil-spill, is directly threatened by southern industrial ways.

Even so, there may be too much to discuss in EARPs. For a government setting up such a process, it may seem an easier route than facing the implications of Inuit political demands squarely in talks on self-government. This, of course, compounds the confusion for Inuit, who may come to expect even more of these forums than they otherwise might.

Environmental assessment in the north is being given inappropriate and ill-fitting tasks. It is being used as a scapegoat for the failure of governments in Canada to determine the futures of the north, to determine the share northerners will have in the exploitation of the wealth under their feet and to accord self-government to a group of citizens on the well-established Canadian model. As Inuit frequently point out, their occupation of the Arctic provides Canada with its claim to vast lands and their inherent riches. Canada should honour that fact and respond by accepting Inuit as citizens with full rights to share in the advantages and opportunities of this country.

In Alaska, Inuit have fought long and hard for an environmental management plan for their coastline and the linked offshore and onshore environments. After many years they have succeeded, with industry and government becoming more co-operative when they better understood Inuit needs (see Anjum 1985). In Greenland, the home rule government has environmental jurisdiction onshore and offshore. There, as in Alaska, Inuit have been prepared to accept development where they feel confident that environmental risks are not too great. The picture of Inuit as opposed to all development is wrong, but interventions before assessment panels are one of the ways they have to talk about their more basic concerns.

The only way environmental assessment can efficiently accomplish its assigned tasks is for it to operate within a framework mutually accepted by Inuit and southerners. That cannot happen as long as Inuit rights to the resources on which their livelihoods depend are put at risk by development plans. It cannot happen as long as Inuit are powerless to affect decisions about their future and given a chance only to express their concerns before a panel. Inuit know that the industrialization of the Arctic will change everything, forever. They cannot be secure in the face of repeated threats, and they cannot be expected to talk only within the neat bounds of bureaucratically defined subject matter. If white people know the specialist distinctions of the south, the Inuit know that what affects the resources and waters affects everything in the north, and they are right. They need their own public institutions and their own experts to help arrive, in co-operation with the federal government, at the limits and standards of development, and they need the jurisdictional clout to manage their own society, which is under the many pressures of industrialization.

In other words, the settlement of Inuit claims and the awarding of political jurisdiction are the only way to make environmental assessment the useful, shared planning tool it could and should be. Environmental assessment is now a tool in the service of industrialization, a public-relations gimmick of government, an alternative to resolving basic policy issues, and a way of hanging onto an absolute jurisdiction over resources that is contrary to Canadian political tradition.

When Inuit are in a stronger position thanks to claims settlements and further self-government, more thinking will be needed. Inuit have not yet addressed these questions, being concerned more with immediate specific decisions than future general processes. Also, there may have been a too-ready belief that negotiating processes would be problem-free if only Inuit were in charge. But experience shows that divisions of view are a fact of life and occur all the time.

In Greenland, the government overcame the wishes and fears of the eastern Greenland and Inuit community of Scoresby Sund, a traditional hunting centre, when approving the Jamieson Land oil exploration program. In Alaska, detailed mapping and subsequent comprehensive coastal-zone planning measures, prepared by Inuit for the North Slope, provide a balanced approach to each area. Although developed in detailed consultation with each village, the planning also brought in the oil companies, which contributed significantly and were pleased with the result. However, Canadian Inuit leaders who have looked closely at the North Slope plan fear that it may lead people to focus too heavily on technical detail and not enough on the "big picture."

In Canada, Inuit do not dispute the federal government's major role in ocean matters. Indeed, they would like government to participate in full and co-ordinated fashion in planning for development. However, they are very uncomfortable about having oil companies coming to them directly and government urging on them a consultation process, as the federal Department of Indian Affairs and Northern Development did in relation to Bent Horn in early 1984. Public authorities must be involved in, and must manage and fund a consultation process. In the Bent Horn case, Inuit refused to accept plane rides and other facilities offered by Panarctic Oils and got the territorial government to offer alternate forms of assistance.

Inuit want community discussion and want to play a role at the community level. However, they also want access to expertise and advice from other levels – national and regional bodies and territorial governments – as shown by the Bent Horn episode. Inuit in that area were subject to many environmental assessment processes over the years, and Inuit in the villages appealed to regional, national, and international Inuit associations for advice. The situation was not, as Panarctic claimed, a case of "outside" organizations interfering.

The reckless way in which government and industry handled the Bent Horn proposal is hard to explain. Coming on the heels of the excellent final report in the Lancaster Sound study, and ignoring the opportunity for consensus among Inuit and other interests, the experience merely proved that government was slow or uninterested in learning. Panarctic's wavering and posturing – an apparent embarrassment to many within the oil industry – have also badly soured the climate for further development. What industry does not seem to have learned is that winning a smile from one's Inuit employees, or winning a small jurisdictional battle, is not the end of the war. Industry and government will meet again and again. If only industry in the north could be as sophisticated as a process-weary public!

In the Lancaster Sound case, an important solidarity has been demonstrated among four communities relatively distant from each other but all significantly involved in the fate of the area. Grouping together interested villages and recognizing them as a collective interest group form an invaluable precedent. However, the major player in the debate has been Pond Inlet, the strongest and largest settlement in the area, and the one most directly affected.

Clearly the mapping of Inuit coastal-zone management decisions would be an enormously useful tool, provided that flexibility was offered in specific situations, so that industry, government, and local Inuit could design their broader future. This is surely an Inuit matter: there is no other public in the Arctic!

A multi-level process, very much in the Canadian tradition, would appear workable and desirable. What Inuit want most of all is a public process: where the pros and cons of projects, and the implications, are discussed and weighed, where people can learn and listen and offer ideas and be given the evidence needed to make up their minds.

CHAPTER SEVEN

Coming of Age: Territorial Review

John Donihee and Heather Myers

The involvement of the government of Northwest Territories (GNWT) in assessing northern resource development projects has changed substantially over the past decade, reflecting general political development in the north. While the territorial government has no direct jurisdiction over marine shipping, its interest and involvement in the review of shipping projects stemmed from its responsibility to its people and for certain elements of the northern environment.

The beginning of formal assessment hearings into environmental and social effects in the north could be traced to Mr. Justice Thomas Berger's Mackenzie Valley Pipeline Inquiry, conducted throughout the Mackenzie Valley in the mid-1970s. At that time, the GNWT's Department of Natural and Cultural Affairs, forerunner of the Department of Renewable Resources, had only one biologist on staff. Today, impact assessment is a significant focus for the department, whose staff has grown to over 200, with a dozen staff members regularly assigned to assessment of development projects, at various stages of the regulatory review and approvals process.

This chapter traces the development of the GNWT's role and capabilities in impact assessment and review – looking principally at formal public reviews of proposals for northern hydrocarbon development and identifying, where possible, marine shipping components. Most of the GNWT's involvement, and indeed that of all government agencies, has been early in the development process – in hearings and reviews. Experience with monitoring, surveillance, enforcement, and management of large-scale projects in the north is limited to the only two approved projects – the Norman Wells Oilfield and Pipeline Projects and, more recently, the more modest Bent Horn Project.

Our purpose is to describe the changing involvement in these project reviews for GNWT officials and elected representatives, by describing

Table 1
Evolution of GNWT Involvement in Hearings, 1975–84

Project	Level of Involvement					Location Where	Evidence	Type of Involvement	
	Politicians	Tech.	Policy	No. People	$			Cross-Exam.	Other
Mackenzie Valley Pipeline Incuiry (Berger) 1975–77	No	Yes	No	no one full-time	–	Yellowknife Inuvik	2 technical witnesses	No	Limited, final argument
Lancaster Sound Drilling EARP 1977–78	No	No	No	1	–	Frobisher Bay Resolute	Informal presentations, observers at hearings	No	Observer on panel
Lancaster Sound Green Paper 1979–82	No	Yes	Yes	1 staff 1 consultant	–	Ottawa	Submission by MED	No	Observer background paper
APP EARP 1980	MED High Arctic MLA	Yes	Yes	2–3	–	Resolute and communities	Submission by MED	No	One GNWT panel member
Norman Wells EARP 1980–1	2 MLAs MED	Yes	Yes	10+	25,000	Yellowknife	2 policy papers, 7 submissions (4 depts)	No questions of challenge permitted	DRR final statement, response to questions
Norman Wells NEB 1981	MED, supported by EC	Yes	Yes (same as EARP)	10–15	50,000	Edmonton Yellowknife	Policy paper, tech evidence (3 depts)	Yes	DRR reponse to questions Final argument on behalf of GNWT
APP NEB 1982–3	Ministers (Energy, DRR) EC approved submission	Yes	Yes	15+	50,000	Resolute Ottawa Frobisher	8 witnesses: 1 policy and 7 technical (5 depts)	Yes	Cross-examined by proponents and other intervenors DRR cross-examined extensively
Beaufort Sea EARP 1981–4	Direction from EC and govt leader	Yes	Yes	35+, extensive involvement	200,000 approx	Yellowknife, Inuvik Resolute Ottawa Aklavik Coppermine Whitehorse	8 submissions by DRR 11 by other depts 30 people (8 depts) Total – 36 participants	Yes (7 cross-examiners)	Opening statement Presentations at community sessions, 5 requests for information Closing statements, final argument

Note: DRR = Department of Renewable Resources; EC = Executive Committee; MED = minister of economic development.

Table 2
Description of Issues Presented by GNWT

Project	GNWT Presentations
Mackenzie Valley Pipeline Inquiry	Dept of Natural and Cultural Affairs – caribou biology; effects on caribou; objectives and activities of GNWT Fish and Wildlife Service Legislative assembly – resolutions passed re: socio-economic, employment, compensation, and environmental concerns; provincial status; settle land claims GNWT final argument – territorial government programs, need for co-ordination of government activities
Lancaster Sound EARP	No submissions
Regional Study APP EARP	Background paper on socio-economic conditions GNWT – training, hiring, northern business development; infrastructure planning; taxation and revenue; territorial energy policy DRR – cumulative environmental effects of disturbance and spills; completion of Lancaster Sound Regional Study; effects on polar bear, caribou, and muskox; lack of mitigative measures
Norman Wells EARP	GNWT – departmental policy and position statements and, later, comprehensive action plans responding to panel's recommendations DRR – need for land use planning, compensation, environmental protection and contingency plans, restoration, means of project regulation
Norman Wells NEB	GNWT – position statement; cross-examination of IPL; departmental witnesses presented evidence DRR – wildlife concerns; similar issues to those presented at EARP; land use planning and proposed mechanisms; work requirements identified for recommendations
APP NEB	GNWT – position statement; 7 GNWT depts made presentations – comments on technical aspects of NEB submission; specific territorial ordinances and territorial v. federal responsibilities; executive council conditions for approval DRR – evidence regarding long-term planning, land use planning, and cumulative assessment; protection of resources and harvesting; information and public involvement requirements; need for mitigation and contingency plans, management structure, and GNWT involvement in monitoring and research; process issues; need for land claims settlement; recommendation and work requirements for each
Beaufort Sea EARP	GNWT – extensive involvement in technical and procedural issues and in hearings and presentation of recommendations DRR – protection of renewable resource economy, preference for phased development; need for northern benefits, mitigation of effects, compensation, land use planning, adequate data and funding; government management and co-ordination of programs; harvesting and wildlife effects; and contingency planning

Note: DRR = Department of Renewable Resources.

the scope of their efforts and by outlining some of the issues addressed. Although there were differences in nature, scope, and timing, among the various public hearings, a clear trend toward greater GNWT partic- ipation is apparent. The results of our review are summarized in Tables 1 and 2.

REVIEW PROCESS FOR THE NORTHERN PROJECTS

The northern project review process is perhaps unique. It is certainly not simple, nor has it ever been clear in advance exactly which steps proponents of a development must take to secure approval.

The ultimate authority for approving most northern development projects is the minister of the federal Department of Indian Affairs and Northern Development (DIAND), although in certain energy develop- ment projects, the National Energy Board (NEB) can become involved. In these cases, a federal cabinet decision will take precedence over that of the minister of DIAND. The complexity of the northern regulatory regime comes as a rude shock to many would-be developers venturing north of 60 degrees for the first time. The variety of committees, re- views, approvals, and permits required to mount an operation rivals that in any southern jurisdiction. Difficulties in clearing these hurdles are compounded by the geographic separation between the operational reality – communities affected in the north – and senior federal policy- and decision-makers in the south. The GNWT's review, policy, and decision-making system is in the north, and its policy and political concerns must be communicated to the south. Little wonder that the "system" can be frustrating, time-consuming, and expensive.

There is little consistency in approach to reviewing different kinds of projects. Although most regulatory licensing and issuing of permits takes place in the north, the screening component of the federal En- vironment Assessment and Review Process (EARP) for different types of projects can take place (pursuant to DIAND policy) in different places. Hydrocarbon developments, major hydro-projects, and highway pro- posals are screened in Ottawa; access roads and trails, mines, and small projects are screened in the north. Developers are often forced to deal with authorities both in Ottawa and Yellowknife (and sometimes Ed- monton and Winnipeg) simultaneously.

The general practice is review in the north by committees (a whole series of them – the Land Use Advisory Committee, the Federal-Ter- ritorial Lands Advisory Committee, the Regional Ocean Dumping Ad- visory Committee, the Technical Advisory Committee to the NWT Water Board, and so on) which advise the decision-making department on regulatory or operational concerns. Policy questions or contentious issues are referred to either regional headquarters, for decentralized departments, or to Ottawa. All of this, of course, limits access to the decision-making process and its accountability to residents of northern communities. Major decisions on developments (to proceed or not to proceed) or on other issues of federal policy are always made in Ottawa.

The GNWT's skills and capabilities in project assessment and review have evolved against this backdrop. Changes to federal review processes can be motivated by national concerns, such as the National Energy Program in the early 1980s and the subsequent Canada Oil and Gas Act, which resulted in reorganization of review and approvals processes for oil and gas development and creation of the Canada Oil and Gas Lands Administration (COGLA). Competition between this new organization and branches of DIAND for control of northern hydrocarbon development back the GNWT's interests and involvement in resource development and decision-making.

The geographical separation of policy- and decision-making from resource development operations and the affected communities reduces meaningful public input, flexibility, and accountability. Opportunities for direct negotiation and creative approaches for affected northerners and proponents in coping with development are lost, making the development process more frustrating and expensive for all who are involved and inhibiting rapid development of the capabilities of northern governments. It has blurred distinctions between the national interest and local and regional interests in decision-making about northern resource development.

HEARINGS

Mackenzie Valley Pipeline Inquiry

Mr. Justice Thomas Berger's mandate in this inquiry was to investigate the social, environmental, and economic impact of a gas pipeline down the Mackenzie Valley. The "Berger Hearings" established the norm for all hearings held since in the north. In time, effort, and expense, his review was even more exhaustive than the Beaufort Environmental Assessment Review Process (BEARP). The inquiry captured the attention of the nation, and Mr. Justice Berger's approach, based on both technical and community hearings, has provided a model for subsequent reviews. The federal assessment process, and even the NEB review of the Arctic Pilot Project (APP), have borrowed from it.

The GNWT played virtually no part in the Berger Hearings, deferring largely, even on wildlife matters, to federal experts. Officials from the Department of Renewable Resources made only two technical presentations before Mr. Justice Berger, after requests by inquiry staff. These presentations concerned potential effects on caribou and ongoing caribou studies, as well as the objectives and activities of the GNWT's Fish and Wildlife Service – primarily, its maintenance of wildlife populations for use by native harvesters.

The speaker of the legislative assembly made a presentation consisting of resolutions passed by the assembly, dealing with socio-economic concerns, employment, compensation to native harvesters for lost wildlife, and adequate provision for environmental protection. Further arguments were presented regarding provincial status for NWT and settlement of native land claims.

The GNWT also called attention to its numerous programs that had been overlooked, or for which duplication was proposed in other agencies' recommendations, and called for co-ordination, rather than duplication, of government management activities. The GNWT felt that it already provided many of the services called for in intervenors' presentations. It believed that it knew best what northern residents wanted and needed, although this view was not necessarily shared by all northerners.

Lancaster Sound Drilling (EARP and Regional Study)

In 1977 and 1978, the Lancaster Sound EARP review was conducted to evaluate a proposal by Norlands to drill an exploratory well in deep water in Lancaster Sound. The GNWT reviewed the technical documents but made no submissions to the review panel. Neither elected officials nor policy staff were involved, but some field staff in the communities provided limited input to the hearings. An observer was appointed to the panel after its vice-chairman commented on the lack of involvement by the GNWT.

The GNWT put more time and effort into the subsequent Lancaster Sound Regional Study. One member of its Baffin Region staff participated, and a consultant was hired to produce a background paper on socio-economic conditions in the region and be more continuously involved.

Arctic Pilot Project (EARP)

The EARP reviewed a proposal involving development of gas-fields on Melville Island, an overland pipeline, liquefaction plant, and shipment of liquefied natural gas (LNG) to markets in the south.

By the time of the hearing in 1979, the GNWT's attitude to these kinds of review had changed, and its involvement was more extensive. The minister of economic development made a presentation about training and hiring, northern business development infrastructure planning, taxation and revenues, and territorial energy policy. Environmental issues raised included the cumulative effects of wildlife disturbances and oil spills, completion of the Lancaster Sound Regional Study before approval of development, polar bear conflicts with humans, and the lack of mitigative measures outlined by the proponent.

Other presentations were made by the member of the legislative assembly (MLA) for the High Arctic, and staff of the Baffin Region, who also worked with the communities and reviewed technical documents. The GNWT was represented on the EARP Panel by its deputy minister of economic development.

Though costs of participation in this hearing were relatively minor, the government's profile was enlarging, as was its determination to participate in these formal project reviews.

Norman Wells (EARP)

Before the APP was reviewed by the NEB, the Norman Wells Pipeline proposal was submitted to the government by ESSO Resources and Interprovincial Pipelines Ltd. During the reviews of this project by EARP and NEB, the GNWT reached new levels of involvement, through its review of the proposal and the preparation necessary for the hearings.

The Norman Wells EARP represented the first substantial effort by the GNWT to publicize its own position. The minister of economic development again appeared with a position statement. Three other MLAS and several deputy ministers, as well as staff members from the departments of Health and Social Services, Local Government, and Renewable Resources, also made submissions. For the first time, technical consultants assisted in the development of submissions. Renewable Resources answered questions and, along with the departments of Economic Development and Local Government, responded to panel recommendations. The minister of renewable resources made a policy statement, and the assistant deputy minister outlined issues for panel review, including land use planning, harvesters' compensation, environmental protection and contingency plans, rehabilitation and restoration of habitats, possible effects from summer maintenance and repair activities, and managing regulation of the project (in a fair amount of detail).

The EARP Panel's recommendations incorporated many of the ones made by the GNWT. The GNWT subsequently developed comprehensive action plans to respond to the EARP recommendations.

This was the first time that a "political consensus" on a "northern position" was achieved in hearings between the GNWT and the aboriginal organizations, namely, the Dene Nation and the Metis Association. These parties agreed to intervene on the basis of a series of mutually acceptable points, which also guided subsequent intervention in the NEB hearings.

Norman Wells (NEB)

The Norman Wells proposal was the first time that the GNWT was involved in the more rigorous, quasi-judicial NEB hearings. However, with experience from the EARP review, the government was well prepared. Nevertheless, some GNWT departments, because of the difference between the two processes and a perception that they had already had their say, participated less in these hearings.

The GNWT participated in both the Edmonton and Yellowknife sessions and prepared and presented technical evidence, as well as cross-examining witnesses and submitting a final statement. The GNWT's position statement was given by the minister of economic development, as part of a strategy plan to play a more active role in review processes. Presentations were made also by the assistant deputy minister and by the superintendent of wildife from Renewable Resources. These dealt with wildlife concerns, and issues similar to those presented at the EARP hearings, as well as land use planning (in more detail).

Although the GNWT intervened in the NEB hearings, the decisions resulting from this review process were made by the federal cabinet. Under the conditional approval eventually granted by federal order-in-council, Interprovincial Pipelines was required to prepare a number of environmental and social reports to satisfy outstanding concerns. Development of terms of reference and evaluation of them were the responsibility of NEB staff alone. Intervenors, including the GNWT, could review these reports but had little influence on their content. Thus, the GNWT, despite its more aggressive role and expanded involvement, was still merely an intervenor and adviser to a federal review process.

Arctic Pilot Project (NEB)

Once again, for the Arctic Pilot Project (APP), the nature and scope of GNWT involvement were expanded. The ministers of renewable resources and of energy and resource development presented a government position paper endorsed by the full executive council. Over fifteen members of various departments presented technical and policy papers. For the first time, outside experts were used as witnesses (personnel from the Canadian Wildlife Service – CWS – prepared evidence regarding seals and marine birds). Participation spanned hearings in Ottawa, Resolute, and Frobisher Bay. Legal counsel representing the GNWT was in continuous attendance at the hearings in Ottawa.

The GNWT's preparation for and response to this hearing were more co-ordinated than ever. In preparation, a working group composed of deputy ministers and staff from seven departments was called together, to outline the government's concerns. Members commented on technical aspects of the proponent's NEB submission and reviewed it with respect to territorial legislation, as well as to distinguish territorial responsibility and concerns from those of federal departments.

The ministers of renewable resources and of energy and resource development presented a list of conditions approved by the executive council for GNWT support of the project. These conditions indicate the range of the GNWT's mandate and its perception of community interests and concerns:

– protect archaeological and historical sites;
– work with communities to minimize negative socio-economic effects;
– work with communities to develop strategies;
– prefer northern businesses where possible;
– conduct a needs assessment and then develop employment and training programs with the GNWT;
– provide energy to accessible communities;
– minimize damage to flora and fauna;
– protect renewable resource harvesting and provide compensation where necessary;
– identify environmental baseline data gaps, monitor effects on renewable resources, and develop appropriate plans and studies, which would involve the Hunters' and Trappers' Associations (HTAs) and the GNWT;
– not affect community resupply;
– develop a project management structure representing affected groups;
– follow territorial health and safety ordinances;
– conform to any environmental management regimes set out by land claims settlements, and
– allow for sharing of resource revenue.

The assistant deputy minister of renewable resources filed evidence on environmental concerns, which included considering any long-term effects resulting from the project in the design of any future hydrocarbon developments; planning land use and cumlatively assessing environmental effects; giving HTAs and native organizations a role in research and development; providing environmental baseline information; setting up a mitigation and contingency plan; creating a project management structure; allowing GNWT participation in monitoring and research; and protecting renewable resources and harvesting opportunities.

The GNWT also raised some questions regarding process. It argued that there should be a new hearing if oil were to be transported instead of LNG; that any major changes to specifications for the LNG tankers would necessitate further public review; that there were no Canadian regulations regarding LNG carriage and safety; that the Terminal Port and Pollution Code (TERMPOL), instead of being voluntary, should be mandatory; and that staff of federal departments should be involved in the NEB hearings. (As a matter of policy, federal departments did not intervene before the NEB, so that the federal Department of Environment did not intend to make a presentation.) The GNWT had to make a request at the federal level to secure the assistance of specialists from the CWS to present evidence on marine wildlife matters. The lack of federal staff involvement in NEB hearings on northern projects reduced the technical quality of these sessions. Provincial-government personnel can deal with technical and scientific issues related to resource and environmental matters, but the GNWT's limited mandate reduces its range of expertise. This was a major factor during the NEB's hearings on the APP because of the large marine component and the division of labour among wildlife management agencies. (The GNWT did not, for example, have scientists expert on marine mammals or birds.)

In response to cross-examination by the Inuit Tapirisat and the Baffin Region Inuit Association, the GNWT identified work required to meet each of its recommendations. In a similar manner, work requirements had also been detailed during the Norman Wells hearings and were used to justify subsequent funding requests made to the federal government for GNWT programs.

The GNWT assisted involvement by aboriginal organizations by helping to fund the Inuit Tapirisat's intervention in the NEB hearings. The minister of energy and resource development also mentioned the GNWT's support for the native organization's argument that approval of the APP could prejudice settlement of the Inuit land claim.

Beaufort Sea (EARP)

The Beaufort Sea EARP review occurred far in advance of a formal project proposal, with the intention that it provide a general form of planning process. Government agencies and residents of the Mackenzie Delta were already experiencing the effects of high levels of exploration activity and were concerned about eventual development of the area.

GNWT participation in the review was actively directed by the exec-

utive council. The government leader made public statements in both
Aklavik and at the final session in Ottawa, where he spoke jointly with
the DIAND minister. Involvement by both policy and technical staff was
extensive, with preparations spanning two years, at a cost of over
$200,000, exclusive of salaries. Government officials participated in all
technical sessions and in several community sessions. Nineteen presentations of evidence involving over thirty staff members were made,
and there was substantial effort put into questioning proponents and
other intervenors.

Government participants were involved in the process from the beginning and commented on the hearings as they evolved, reviewed the
proponents' impact statement and submitted a deficiency statement,
made information requests, and delivered a final argument. This latter
consisted of extensive recommendations, each with a detailed rationale.
Many of the environmental concerns important to Renewable Resources
were addressed by the panel in its final report.

Environmental evidence was presented by the deputy minister and
seven staff members of Renewable Resources. The department addressed the need to protect the renewable resource economy, the de-
sirability of phased development, the need for northern benefits, and
the department's interest in helping to mitigate harmful effects and in
ensuring compensation for native resource harvesters affected by development. Staff members gave evidence about government management capabilities, environmental management, the Renewable Resource
Harvester's Compensation Policy, and the need for land use planning
and for adequate funding and co-ordination of monitoring activities.
Experience with responding to, and management of, the Norman Wells
pipeline project were still fresh, and the presentations gave substantial
detail regarding problems experienced with that project. Other issues
were dealt with by evidence on harvesting changes related to industrial
impact, wildlife habitat management, effects on ungulates and polar
bears, contingency planning and contaminant control, and bear/human
conflicts.

Specific marine issues revolved around a preference for a pipeline
rather than ship transport of oil and the studies needed if ships were to
be used. It was recommended that these studies including testing of the
effects of ship tracks on native hunters' movements, the effects of year-
round shipping on ice regimes and marine mammals, oil-spill clean-up
capabilities, polar bear avoidance of oil spills, and the choice of tanker
routes – almost the same issues as those raised during the Lancaster
Sound and APP hearings. Progress toward resolving them had been slow.

The attitude of the residents of the eastern arctic region had changed over this time. During Lancaster Sound and APP deliberations, they were adamantly opposed to either LNG or oil shipment. The GNWT also opposed shipment of oil and made that point in the Beaufort Sea hearings, stating that it preferred transport by pipeline of oil and gas. Reservations remain, in the north, regarding ship transport of oil and gas through the Northwest Passage, but despite these concerns, when the Bent Horn Project was proposed, residents were willing to accept it as a pilot project because of its small size and the phased approach adopted. Panarctic's performance on the Bent Horn project will test the acceptability of tanker transport to Inuit residents of the Lancaster Sound region.

AN EXPERIENCE OUTSIDE HEARINGS: BENT HORN

The Bent Horn Project proposal involved, as a pilot project, drilling and development of a small oil-field on Cameron Island and tanker transport of 100,000 barrels of oil south during the summer shipping season. There was no formal public review. The project received federal review through a variety of committees, and in the early stages the GNWT was not consulted in any consistent manner. The GNWT was left to respond, through bureaucratic channels, to the federal agencies involved, including DIAND and the Canada Oil and Gas Lands Administration. The GNWT accepted the project on condition of establishment of an environmental protection plan, completion of a compensation plan, and revisions to oil-spill contingency plans, among others. Federal officials did not consider these matters seriously until the GNWT was encouraged to participate at the political level, by DIAND's new minister, David Crombie.

THE FUTURE

Despite the evolution of the GNWT's ability to respond to project proposals and regulatory reviews, and its ability to contribute credible suggestions designed to reduce harmful effects and to improve environmental management, acceptance of the GNWT's input still depends on a consultative system that is essentially colonial in attitude and conduct and not necessarily consistently applied. A trend is discernible, however. The GNWT has played a progressively more assertive role in project assessment hearings and reviews. It is better able to articulate a position and to produce technical submissions to support political goals. Its departments now possess the experience and core of capability to deal with hearings and reviews and also have enough experience with environmental impact assessment to analyse seriously the benefits of past participation in federal processes when acting as intervenors. The GNWT deserves a more significant role – on the basis of its proved contribution and jurisdiction.

There should be more local involvement in decisions on northern development, with local residents able to influence decisions that may affect their way of life. The recent interest in joint ventures and development agreements in the Mackenzie Valley represents a move in this direction – the expression of a desire to take part in development planning and to share in the benefits of non-renewable resource development.

Participants should seek to resolve some of the recurring issues in northern project review. Many policy and technical concerns raised regarding hydrocarbon development and marine transport in the High Arctic over the past decade remain largely unanswered. Issues raised during the Lancaster Sound Review were still being raised in the Beaufort Sea hearing, and satisfactory answers were still not available.

"Design" of a northern review process is pressing, particularly given the general lack of enthusiasm for another EARP. The time has come for northern residents and their government to design their own project review and assessment process. One hopes that it will incorporate some innovative approaches, emphasizing the involvement of affected parties in problem-solving and decision-making. This process must seek to resolve fundamental questions regarding the effects of development on the northern environment and society, and it must move toward giving northern people and their governments real control over these developments.

POSTSCRIPT

In September 1988, Prime Minister Brian Mulroney and Dennis Patterson, government leader in the NWT, signed an agreement-in-principle for negotiation of a Northern Accord. This accord, when finalized, will transfer legislative responsibilities for oil and gas management on land to the GNWT. Responsibilities for offshore oil and gas would be shared. The agreement in principle recognizes and protects aboriginal rights and land claims settlements. It provides for the revenue-sharing from northern hydrocarbon development by beneficiaries of land claims settlements and the GNWT. This accord signals a potential new role for northerners in decision-making on resource development.

CHAPTER EIGHT

The Federal *EARP* Experience
David W.I. Marshall

Residents, bureaucrats, and industry look at the north in different ways. Peter Aglak of Pond Inlet observed: "If Lancaster Sound would be used as a shipping route all year round it is difficult to say what effects it will have on the sea mammals. If something should happen to the animals or they move it would be hard for the Inuit to live." Martin Barnett, Department of Indian Affairs and Northern Development (DIAND), commented on the Arctic Pilot Project (APP): "The project offers a rare opportunity to investigate a relatively pristine Arctic environment then disturb it in a controlled fashion and measure how its elements respond. Such information will be valuable is assessing the impacts of future northern developments and instituting useful mitigative measures." Menno Homan of the APP said: "We recognize that there are some significant knowledge gaps resulting from the fact that there is no precedent for year-round ice-breaking with such large ships in the Arctic. As a result we have taken the approach of identifying the major concerns, making a commitment to study these concerns as the project begins, and ensuring that design options exist if problems develop. The answers to some of the questions can only be obtained after year-round ice-breaking is in operation."

Canadians have been moving into their arctic frontier with increasing vigour and interest since the last century. Daring explorers first began casting covetous eyes on the Canadian Arctic in the sixteenth century, in their relentless pursuit of a Northwest Passage to India and China for much-sought-after riches. However, not until oil was discovered in 1968 at Prudhoe Bay, Alasaka, did people become interested in commercial shipping in the Canadian Arctic. Before that time the fur trade was giving way to mining as the premier industry in Canada's north. First gold miners, then copper, lead, and zinc mining companies went north of the sixtieth parallel to seek their fortunes.

In the past decade, energy resources have been the Arctic's most powerful economic lure. Oil and gas have been found in the High Arctic and the western Arctic, and now the petroleum industry is seriously contemplating ways and means of taking this new-found wealth to market. Shipping is one of the main transport options.

This shipping activity would take place in a region that is among the last economic, scientific, and engineering frontiers of the world, a region that presents extraordinary challenges to humankind. For example, it has a sensitive ecology that could be easily affected if proper care and management are not exercised by all participants. Uncontrolled industrial activity could thrust unproven technology on an environment that is far from being fully understood in a biophysical sense. This environment is clearly vulnerable, due to severe climatic limitations and the sensitivity of its biological components, which experience dramatic population fluctuations and slow growth rates.

Before year-round arctic shipping came under serious consideration by the oil and gas industry, the government of Canada recognized that a mechanism was needed to help resolve conflicts over major development projects such as marine transportation of arctic energy. Canada's Environmental Assessment and Review Process (EARP) is one such mechanism that has been established to ensure the identification of key issues associated with arctic shipping and the responsible integration of these issues into sound environmental management and planning.

DIAND is responsible for ensuring that potential environmental and directly related socio-economic impacts of arctic marine transportation activities are taken into account early in the planning process, that is, before irrevocable decisions are made. In the north, Lancaster Sound drilling, APP, ore shipments from the lead/zinc mine on Little Cornwallis Island, Beaufort Sea oil tankers, and the Bent Horn tanker proposal have all been subject to various stages of EARP.

This chapter explores the EARP experience associated with decision-making about arctic marine shipping and attempts to assess some of the lessons to date. It also points out the value of EARP in identifying the more significant scientific issues associated with arctic marine transport and what government and industry need to do in both the short and long term, to resolve outstanding scientific questions.

EARP

In 1973, the government of Canada established the federal Environmental Assessment and Review Process (EARP). This process was established by cabinet order to ensure that the federal government's activities (or initiatives involving the private sector over which the federal government has a decision-making responsibility) are examined for potential environmental effects early in the project planning process. EARP is predicated on self-assessment: the main decision-making government agency for the project conducts an internal preliminary assessment of a proposed development.

In many instances, environmental effects are not expected to be significant. Such projects are allowed to proceed while sound environmental design principles and mitigative measures are incorporated into their implementation. However, where the decision-making agency sees potential for significant environmental harm, the proposal must be referred to the minister of the environment for independent review by an EARP panel. The review is organized with the financial, administrative, and technical support of the Federal Environmental Assessment Review Office (FEARO), which administers EARP. FEARO reports directly to Environment Canada. Historically, only a few projects have been referred for independent public review by an EARP panel.

Most members of an EARP panel are generally drawn from outside the federal public service. An attempt is made to find people with expertise that can contribute to full understanding of one or more aspects of the proposal under review and a background that will enable them to view the proposal from a broad perspective. Moreover, panel members must have no real or apparent vested interest in the proposal. Normally, panel members are supported by a number of independent technical specialists, hired with the administrative assistance of FEARO to review specific elements of the proposal. These specialists are asked to provide impartial expert opinions and scientific data on certain issues and to raise other issues that might be overlooked during the review process.

Upon completion of the public review, the panel prepares and provides a report to the minister of the environment and to the minister responsible for the agency proposing to undertake or authorize the project. This report, which contains the panel's recommendations, is made public. The panel's recommendations are advisory only and not binding. The minister of the agency responsible is expected to provide the public with a response to the panel report, before terms and conditions under which the project can or might proceed are finally established.

On 21 June 1984, EARP was strengthened and made more comprehensive by order-in-council. This revised mandate clarifies departmental obligations, streamlines procedures, and makes information even more accessible to the public.

EARP is neither a decision-making nor a judicial process. It is a mechanism for considering all aspects of a proposed development that could pose significant environmental threats. It allows the developer's plans to be reviewed by both experts and generalists, by organized vested interests, and by ordinary citizens. EARP is a way of applying scientific analysis to contentious issues, while providing a forum for public expression of views. The process allows proposals to be considered for their environmental implications, without prejudice as to land and resource ownership.

Not surprisingly, environmental impact assessment (EIA), which is a major part of EARP, has given rise to many concerns, such as cost, uncertainty as to data and research requirements, and delays on project decisions. These issues, and suggested recommendations for their resolution, are discussed later in this chapter.

ARCTIC ENERGY SHIPPING PROPOSALS

Ice conditions along the arctic route vary considerably from year to year, with first-year ice ranging from 1.6 m in Davis Strait and Baffin Bay to 2.2 m in Viscount Melville Sound. The action of winds and surface currents on the ice cover creates large numbers of pressure ridges, and the ice pack of Davis Strait has many multi-year floes, thousands of icebergs, and bergy sections. Furthermore, the shipping route through Parry Channel is a significant migratory route for seabirds and marine mammals. Mammals enter Lancaster Sound when the fast-ice breaks up and then migrate westward to Barrow Strait and elsewhere. During late summer and fall, some species of marine mammals and seabirds are widely dispersed in and along the edge of the pack ice in Davis Strait and in western Greenland waters.

The Arctic's severe environmental conditions pose major challenges to the design and operation of marine transportation systems. In addition, human activity may effect the atmosphere, ice, and sea-state regimes. These developments may cause the following effects: change ice stability and the location of the landfast ice edge, create artificial leads, produce ice fog, redistribute ringed seal pupping lairs, alter polar bear range, and change other related biological and physical activities. Six major environment issues associated with year-round shipment of LNG or oil in the Arctic by ice-breaking tankers identified through the APP and Beaufort Sea reviews are discussed in greater detail as follows.

The Arctic Pilot Project

The Arctic Pilot Project (APP) was the first proposal to operate year-round shipping in the Arctic. It was the first proposal involving energy transportation that required ice-breaking through heavy ice during the course of normal operations. The APP tankers were to be class 7 and would have a liquefied natural gas (LNG) cargo capacity of 140,000 cu m. There were to be two tankers initially, and each vessel was to make fifteen round trips per year. The shipping route would traverse Parry Channel (Lancaster Sound, Barrow Strait, and Viscount Melville Sound), Baffin Bay, and Davis Strait.

The APP had two central objectives: to produce natural gas from the High Arctic on a relatively minor yet commercial scale and to demonstrate the viability of year-round marine transport from the Canadian Arctic. APP's projected cost was $1.5 billion (1980 dollars), which included a $220–million, twenty-year research and development (R&D) program. This budget figure demonstrated the sponsors' commitment to develop systems to ensure secure, reliable, and environmentally safe operations. As the research program was designed to be spread over twenty years, there was adequate time to conduct studies that required sustained efforts. The sponsors believed that with a strong and broadly defined R&D program the APP would add greatly to Canada's northern transportation technology. They were also convinced that the APP's pilot commercial new technologies – in navigation, oceanography, environmental analysis, arctic engineering, management of emergencies, and vessel design and operation – could form an integral part of successful vessel-based bulk transport from the Arctic.

Concerns were raised about the APP's possible effect on wildlife behaviour, ice regimes, and aboriginal affairs. Petro-Canada, a crown corporation, and DIAND recognized these concerns and requested the minister of the environment to appoint the APP EARP in November 1977, to assess the potential environmental effects of the northern component of the APP.

The panel examined all aspects of the proposal and concluded that the shipping component would be environmentally acceptable if shipping routes avoided environmentally sensitive areas in Parry Channel and if full advantage were taken of the opportunity offered by the proposal's "pilot" nature to monitor and research the effects of year-round shipping in the Arctic, so that this knowledge could inform subsequent shipping proposals. The panel added that this goal could be achieved only if a control authority was established by the federal

minister of transport, to monitor ship movements and enforce good
seamanship and appropriate environmental regulations. The panel stated:
"Without further research on marine mammals, guided by the advice
of Inuit, and of government scientists, and without a monitoring and
control mechanism for the selection of the shipping routes, the Panel
is unable to recommend that the project is environmentally acceptable."

The panel recommended that the federal departments of Fisheries and
Oceans and of Environment form an advisory committee, to include
representation from Inuit, the APP, and appropriate government agencies. The committee's initial task would be to recommend and approve
studies on integrating biological information into the route-selection
process.

The basic message of the panel's recommendations was one of experimental management: scientific information obtained would be used
to guide development and management of both this project and future
arctic shipping interests. Scientific information would then continue to
be an integral component of the decision-making process. Many people
believed this approach to be responsible: the panel could have taken an
easier course and suggested that no decision be made until further research on wildlife activity and ice regimes was completed. Unfortunately, the basic biological and physical research was not completed on
a full test-case basis (although subsequent research by Dome on its own
Beaufort Sea oil tanker proposal did pursue some of the questions
raised), since no markets for the gas became available. The project was
eventually abandoned.

Nevertheless, the panel did make a significant contribution to future
decision-making on arctic shipping. Its recommendations on advancing
the biological component of an integrated route analysis for future
shipping and its concept of a control authority had much broader applications. Some people believed realistic estimates of as many as 30 to
50 ships carrying all types of cargo and making up to 1,000 annual
transits by the year 2000. The panel took this possibility into account
by stating its concern about potential environmental effects of large-
scale year-round shipping in the Arctic. It recognized the obvious need
for further research, and members were adamant that any shipping
proposal to pioneer year-round shipping be small-scale and include
effective monitoring to assess and mitigate environmental effects.

The proponent for the APP had described the proposal as a "pilot"
project, to prove the technical and economic feasibility of year-round
ship delivery of natural gas (and possibly other sources of energy) from
the Arctic. The APP Panel had agreed with this rationale "in the sense
that it [the APP] would pioneer year-round Arctic marine transportation

and develop in Canada greater Arctic expertise within industry and government."

Beaufort Sea Oil Tankers

At the same time as the APP was being considered as a possible option for moving LNG from the Canadian Arctic, Dome Petroleum was making some initial oil discoveries in the Beaufort Sea Region. Enough evidence of oil was found to begin contemplating moving the oil out of the deep-water oil-fields by tankers. There are more than ninety seismic irregularities in the offshore region, of which one-quarter may be capable of commercial production. Since 1976 Dome Petroleum has participated in drilling twenty-eight wells in the Beaufort Sea and has had ten discoveries, of which five have been primarily oil, and five primarily gas.

DIAND believed that this discovery phase marked the appropriate point at which to review the proposal, while major project choices had still to be made. Thus DIAND referred it to the minister of the environment for an EARP panel review in July 1980. Before irreversible decisions were taken and project designs settled, important aspects of development in the Beaufort Sea could be considered holistically, rather than on a piecemeal, project-by-project basis.

All seven members of the Beaufort (Sea) EARP (BEARP) Panel, including the chair, were from outside the public service – the first entirely non-government EARP. Careful recruiting was undertaken to choose panel members who could represent a number of important disciplines and backgrounds, in addition to having intimate knowledge of the Arctic and its people. In a review such as this one, many highly technical and complex scientific issues were addressed. Consequently, the panel was supported by sixteen technical specialists, conversant on a variety of subjects, ranging from physical and biological oceanography and marine design to the effects of underwater sound on marine mammals.

The panel's most significant conclusion was that production of oil and gas and transport from the Beaufort Sea were acceptable, provided that activities associated with development were carefully controlled, were initially on a small scale, were phased in gradually, and were managed with extensive regional participation. The panel believed this course of action preferable, since any harmful effects would be more manageable and potential benefits would be enhanced. In its report, the panel outlined ten conditions to ensure the environmental and socio-economic acceptability of arctic small-scale oil production and transportation.

The panel defined small-scale development as being production of about 100,000 barrels of oil per day and recommended initial transport through a 400-mm (16-in) overland pipeline. It also recognized that the first major commercial oil discovery in the Beaufort Sea could be found in a deep-water location. Panel members were aware that if that were to happen, the oil industry would probably be more attracted to tankers for transporting the oil. They believed that a great deal of information was still needed to prove that tanker transport in the Arctic was within acceptable environmental and socio-economic limits. Members also believed that government preparation was not at a stage to ensure safe transport. Therefore, the panel recommended tanker transport of oil only after specific research and preparation and after tankers were tested in a demonstration project; there was adequate time to fill information gaps and complete preparation before the need for year-round transport of oil was confirmed, provided that industry and government moved quickly.

The panel decided that early research should focus on collection of baseline information on the distribution and normal behaviour of wildlife, on hearing sensitivity and community processes of marine mammals, and on natural variations in ice regime. Preparation activity should include government support, such as provision of systems to detect weather, ice, and hazards; hydrographic charts and communication systems; and government and industry oil-spill contingency plans.

Although the panel accepted the comprehensive design and performance objectives proposed by Dome Petroleum for the Beaufort Sea oil tankers, the panel and Dome agreed that Canadian Coast Guard inspections and sea trials were an integral part of ensuring that vessels met all the conditions necessary for their intended use. The panel suggested that the tankers be tested while baseline data were being collected, but in an area with ice-covered waters of similar ice thickness. The panel believed that the results of all these activities could be useful in formulating conditions for tanker use, which could be adjusted according to a tanker's performance over time. These conditions included rerouting of ships, altering schedules to avoid critical times or areas of biological activity, changes to ship speed, and cessation of tanker traffic.

Panel members knew that a full testing program could not begin until tankers actually operated in the Arctic. As a result, it recommended that two vessels undertake the demonstration phase, in order to provide an alternative in case one tanker experienced operational difficulties. In this phase, the panel suggested comprehensive monitoring, to evaluate the effects of the oil tankers on the distribution and behaviour of wildlife and on ice regimes. If this phase indicated significant adverse effects, government decision-makers would then have sound reasons for changing conditions for use of tankers. This situation is similar to that proposed for the two APP tankers: a test, or "pilot" case to prove the feasibility of year-round marine transport in Canada's Arctic.

The Beaufort Sea Panel also made a number of general recommendations about research and monitoring and also identified two major areas for further research: long-term research into basic physical and biological processes of the arctic environment, and primary research to complete necessary baseline data. Such work would provide data on which to develop effective monitoring and mitigation programs better to assess and soften the effects of development. The panel added that scientific research should be focused on the most efficient and productive activities associated with development proposals to avoid dissipation of scientific effort on too broad a research program.

Tanker Integrity

The oil companies have had many years of experience in the Beaufort Sea with ice-reinforced drill ships, ice-class supply vessels, and ice-breakers such as *Robert Lemeur* and *Kigoriak*. The information gained is being used by Dome and others to help design ships for the future.

The proponents of LNG tankers and of Beaufort Sea oil tankers have committed themselves to "overdesigning" the tankers and employing navigational aids at a standard considerably in excess of the minimum requirements under the Arctic Waters Pollution Prevention Act. For example, the design of arctic tankers would include many safety features not found on conventional tankers. Tankers would have a double hull and a double bottom, which would reduce the risk of a spill in the event of an accident. Adverse ice conditions would more probably slow the tankers' progress than damage the ships. While serious damage is possible, hazards would be reduced by use of an adequate ice- and weather-information system. The possibility of the APP tanker "bumping" into something when moving at higher speeds in the summer season was not raised by the APP Panel but did prove a major concern with the proposed Beaufort Sea oil tanker, possibly because oil created more concern than the LNG of the APP.

Variation in Ice Regime

It has been assumed that regular ship movements through an area may alter ice patterns, causing changes to the ice regime (such as a delay in the formation of ice in Lancaster Sound in the fall). Such changes to the ice regime, if they did occur, might measurably change the local climate and the distribution and abundance of wildlife species in the area. Native people are concerned that whales could get caught in artificial leads. Proponents of the APP and of Beaufort Sea oil tankers believed that changes would be small and would be masked by natural variations in the ice cover. However, the effects of ship traffic on ice break-up and the stability of ice-floe edges are still not fully understood and could become a problem, especially with a significant increase in ship traffic. Both panels recommended further basic research and long-term monitoring, to determine the effects of ice-breaking on ice regimes.

Tanker Accidents and Oil-Spills

The LNG and the Beaufort Sea oil tankers pose somewhat different problems regarding shipping accidents. With the LNG tanker, the main concern is the possibility of an explosion, which would produce intense heat in a localized area. This effect would probably be limited to an area within an 11-km radius of the ship. With the oil tanker, the main concern is the threat of an oil-spill.

Dome Petroleum and the Beaufort Sea Panel thought such a spill extremely unlikely, because of the design of the tanker. Nevertheless, the possibility of a spill was probably the greatest concern raised at public hearings, particularly for people who depend on harvesting of renewable resources for their food or income.

Studies of oil-spills from conventional tankers in temperate waters have indicated that most tanker accidents have involved ships not owned by the major oil companies. The majority of these accidents were caused partly by human error. In some cases, extreme conditions such as high seas or fog contributed to accidents. Human errors have been the result of inadequate training, flouting of the law, breakdowns in communication or authority in vessel operation, honest mistakes, drug- or alcohol-induced accidents, and errors in equipment design.

The Beaufort Sea Panel agreed fully that human error was the most likely cause of a spill. This problem, therefore, was seen to require continuous attention, including special efforts to ensure that employers were aware of the risk of an oil-spill, were trained and supervised in the handling of equipment, and were fully involved in the development and testing of contingency plans to prevent and control spills.

Other concerns about oil-spills centred on the ability to detect and clean up spills that occur under moving sea ice or along the entire proposed tanker route under extreme weather conditions or winter darkness. Determination of shoreline and environmental conditions along the route would assist in planning the location of counter-measure equipment. In addition, such equipment might be installed on board the tankers.

Sound from Vessels

A major issue arising from both proposals was the effect on marine mammals of underwater sound produced by vessels. There is very little information available on this issue because of the difficulty in conducting controlled field experiments and collecting the information necessary to determine if underwater sound is really a problem. For example, little is known about various marine mammals' sensitivity to a range of frequencies at various magnitudes. At present, we know little about communication in marine mammals, echo-location processes, and the potential effects of increased background sound, which might mask these phenomena. In addition, possible changes to population size or distribution of certain mammals caused by chronic stress from under-water sound are relatively unknown.

In some areas, such as the Saguenay and Churchill rivers, beluga whales appear to have learned to adapt to the noise of ships' engines, though over what period of time is uncertain. Both the APP and Beaufort Sea panels recommended more research on the effects of underwater sound. Much of this research would be purely experimental, and it was suggested that the proposed two-tanker demonstration phase would allow for such direct experimentation and observation.

Compensation

Many northerners have stated that they should be compensated for the loss of wildlife harvesting which threatens their livelihood, through either reduced access to wildlife or smaller animal populations. Compensation is an extremely important issue. Since several arctic marine proposals may be turned into realities within the next decade, planning for compensation should begin immediately: the need may arise soon after any shipments of oil or LNG have begun. Specific operators may not be identifiable in some cases, and it is therefore crucial to have clear policies in place before any incidents occur, in order to avoid lengthy legal claims. Both panels recognized this situation and recommended strongly that an acceptable compensation plan be a condition attached to any project approval for the north, with such a policy being developed in full consultation with northerners. (For the Bent Horn Project, a comprehensive compensation plan formed an essential part of the operating conditions.)

Should Alaskan oil eventually be trans-shipped through Canadian waters in vessels of foreign registry, claims in such cases ought to follow an established process and hence avoid lengthy jurisdictional fights and appeals to foreign courts. Since Canadian jurisdiction over the

Northwest Passage is not universally accepted, trans-shipment of dangerous materials should require binding commitments from any vessel operators, regardless of the vessel's registry or flag, to a formal, comprehensive compensation program. Such a strategy protects the interests of residents and enhances Canadian sovereignty and control in the region.

Future Action

Immediate answers to many of the above problems were not given by the APP or Beaufort Sea panel. Although information provided by proponents of both proposals indicated that many of the concerns were no longer valid, both panels recognized that more knowledge was needed, in order to predict more accurately the possible effects of arctic marine transport on the physical and biological environment. It was pointed out that unknowns and uncertainties would complicate management of marine transport, particularly where new technology affects an incompletely known environment. Well-conceived research, intelligence, and monitoring programs should allow for timely responses to unforeseen effects. Present inadequate knowledge of arctic ecosystems, particularly inability to make accurate predictions, necessitates an active research program. As a result, both panels stressed long-term research, better understanding of the physical and biological processes in the Arctic, and demonstration-type projects, to test scientific theories experimentally.

Understanding arctic marine processes that influence ocean circulation, heat exchange, and the dynamics of ice formation and movement is essential for arctic shipping. The general oceanographic regime helps determine the distribution of marine mammals, productivity, and year-to-year variability. Information on ocean and ice dynamics contributes to trajectory modelling and contingency planning for oil-spills to selection of optimum shipping routes, and to indentification of environmentally sensitive areas. Physical oceanographic information is essential for designing ships. A particular priority is the acquisition of physical oceanographic and ice information for development and operation of real-time environmental information and prediction services, such as reports on northern weather and ice-conditions used by vessel operators. Arctic heat-budget and ocean-climate research will reveal the physical conditions that will prevail over the lifetimes of proposed developments.

Knowledge and understanding of marine ecosystems are fundamental to environmental impact assessment. To date, part of the research effort has been directed at the most neglected areas of knowledge, such as the general life history and ecology of certain arctic organisms, rather than at site-specific studies in regions likely to be developed industrially. One of the few exceptions was the research carried out by government and industry in the Beaufort Sea region in the 1970s. The logistical constraints of ship-based operations have limited most arctic marine ecology studies to the waters of the eastern Arctic and to the ice-free season. These studies must now be broadened to include other channels along the proposed tanker route and, where possible, activities during other parts of the year.

Recommendations for each new proposal on arctic marine transportation must rely on short-term applied research and the personal opinions of many individuals. Governments have recently cut funds devoted to long-term basic research. Both panels agreed that if greater attention and funds are not applied to long-term basic research, decision-makers will be forced to draw conclusions about many environmental issues without complete or adequate knowledge.

No simple approach will establish in advance and with certainty the effects of arctic shipping on the marine and physical environment. Therefore, government and industry need to establish a continuing program of research and monitoring to assess effects and to adjust industrial activities and management strategies accordingly.

SCIENCE AND DECISION-MAKING

Decision-making requires deliberate and purposeful examination of situations in need of action, rather than reliance on strictly intuitive investigation. It also requires an analysis of possible alternatives for action and their accompanying consequences, in terms of appropriateness of actions and risks involved. The last phase of this process requires the choice of an appropriate course of action out of the various alternatives analysed.

In environmental impact assessment, decision-making involves critical thinking and requires scientists to examine purposefully and analyse existing biological, physical, and chemical processes to make valid predictions on loss or change caused by external factors.

A realistic role for science in decision-making is now more important than ever. However, judgments must be placed on the quality and quantity of information obtained through science at various stages of making decisions. Because environmental assessment requires decisions based also on subjective judgments involving values, feelings, beliefs, and prejudices, public involvement is an important component. Canadians, especially northerners, are well informed, and only the most parochial decision-maker would fail to tap the knowledge and values people possess about the Arctic and its resources. Therefore, the social, economic, and biophysical character of a region should not be modified without the consent of the majority of its people.

This principle poses an interesting delimma for EARP panel reviews and similar assessment bodies. How does a review process become more streamlined and efficient and at the same time adequately address all relevant public concerns? In some cases, especially when specific target dates are given as part of terms of reference, these two objectives are diametrically opposed.

In some areas the nature, role, and function of the environment are still not fully and adequately appreciated in the formation of economic decisions and activities. In such situations, economic decision-makers tend to treat the environment as something to be considered after economic decisions have been made. The human and financial costs from this approach are already great and could become greater. Economists must begin to assimilate the more integrated, comprehensive view that science brings to the understanding of the natural world and the place of human societies and economies within it.

In theory and in practice, economics focuses on the benefit and costs of particular environmental resources available for exploitation. The scientific understanding of the environment takes an opposing viewpoint. Basically, science sees the environment as a complex, interactive system whose processes support humans and all other living organisms. In any pursuit of economic renewal, the ultimate goal is not growth for growth's sake but a better quality of life for present and future people. Sound environmental and conservation practices make good economic sense at all times.

With this in mind, let us assess the current integration of science and environmental impact assessment (EIA) into the decision-making process and discuss ways of making this integration more efficient. While the following discussion will use examples from the two EARP reviews dealing with arctic marine transportation, similar problems of integration have also appeared recently in most other EIA reviews.

The preceding discussion of the APP and Beaufort Sea review processes and their recommendations indicates a significant involvement of science in EARP and in EIA in general. The panels have aided both types of decision-makers: those responsible for deciding "whether," "where," or "how" a proposal should proceed, and those administering EARP or EIA. This section discusses some of the most important lessons learned from the two EARP panel reviews.

Financial Commitment

The EARP experience with the two arctic marine proposals has given those federal and territorial government agencies directly involved in project decisions for the north some indication of the nature, extent, and location of data deficiencies and has provided some guidance on the direction science must take to fill them. The EARP experience has also shown the importance of high-level commitments to provide the financial and human resources for such research. If there is a shortage of funds, why were incentives such as PIP (Petroleum Incentive Program) grants and tax incentives provided to enable the oil industry to enter frontier areas, if government was not willing to protect ecological and human resources in the event that full-scale production was allowed to proceed? The Beaufort Sea Panel recommendations, for example, may appear costly. However, most of these recommendations must be implemented at some point, and with some the sooner the better, to ensure small-scale development that is environmentally and socially acceptable.

Research

Some may argue that full-scale arctic marine transportation is still a long way off. Coupled with an obvious preference to transport oil and gas by another alternative, such as an overland pipeline, this argument appears to justify ignoring or delaying research and preparation on marine transportation. For example, government funding cuts in the Northern Oil and Gas Action Program (NOGAP) affected primarily research and preparation in marine transportation. However, other government agencies, since the release of the Beaufort Sea report, have given out contracts totaling over $3 million to private consultants, to conduct work in such areas as oil-spill clean-up and damage costs, ice accretion related to marine transportation, behaviour of oil in ice, development of a NAVSTAR airborne dynamic position system, databases, and modeling for arctic marine navigation, long-term effects on the bowhead whale, development of a remote-control communication system for ship-to-shore applications, ice regime (ice observation and trafficability) studies, a shipboard ice navigation system, and arctic-tanker loading and mooring. A relatively expensive contract ($1.2 million) was issued for an over-the-ice hydrographic survey of Arnott Strait and the Queen Elizabeth Islands.

Sooner or later a major oil discovery will be made in deep-water areas of the Canadian or American Beaufort Sea or the High Arctic. Then serious and expeditious examination of marine shipping will become urgent. Two EARP panels, after examining two arctic shipping proposals, have concluded that this mode of transport can be seriously considered provided that certain terms and conditions are met. Now is the time to begin research, so that it can be done in a comprehensive way before Canadians are rushed into it.

Government Management

In a more generic way, the EARP experiences have demonstrated the importance of government management in EIA. In the Beaufort Sea case, the panel early requested statements from a number of federal and territorial agencies on how their current and future policies and programs would affect or be affected by Beaufort Sea developments. Some of these agencies, such as Transport Canada, Environment Canada, and Fisheries and Oceans (Canada), have a significant interest in the marine transport component of the Beaufort Sea proposal. Government management thus became an important element in the panel's review process and had a place on the agenda at the public hearings. The Beaufort Sea Panel devoted an entire chapter to this subject in its report, with the majority of the chapter's recommendations directed to those activities in research and development where the government knowledge base was considered inadequate and to government management programs requiring specific action.

The panel's recommendation on government management was that federal and territorial officials begin immediately to prepare socio-economic infrastructure to deal with all aspects of northern development, with the goal of strengthening local management. The APP Panel recommended establishment of an Arctic Shipping Authority with a general mandate.

Effectiveness and Efficiency

For decision-makers involved in improving the scientific component of EARP and EIA in general, the APP and Beaufort Sea experiences have provided some valuable lessons. Perhaps the most important, and one that continues to plague EIA review processes, is the acquisition and analysis of relevant data. Recently concerns have been expressed that EIA processes are too open-ended and include too many nonessential issues. It appears that developers support impact assessments, as long as information demands are reasonable, duplication is avoided, and reviews are carried out efficiently.

Major EIAs such as the APP and Beaufort Sea are costly for all participants, and everyone therefore wishes an efficient process. That poses an interesting dilemma when a vast amount of information is involved.

Many people believe that more documentation is asked for and received in EIAs than is required to resolve fundamental issues. This may be the case, but where does one draw the line? What is valuable, and what is of no consequence?

There are few things more subjective in life than values. What is seen as important by one person will be seen as something quite different by another. This can be true even among professionals with similar backgrounds. Various factors, such as cultural setting, work situation, and personal experiences, all help to create very different viewpoints. Scientists offering professional advice may disagree with each other, usually favouring their own disciplines and fields of interest. In most cases, they can reach some measure of compromise. For example, the Beaufort Sea hearings dealt with a complex and esoteric subject that was consuming too much time and arousing conflicting views from people with similar technical backgrounds. Scientists representing the proponents, government, and interest groups met under the auspices of the panel's secretariat to reach agreement.

The public affected may disagree with experts, sometimes because of what scientists call misconceptions. Many people are thoroughly sceptical of assurances given by scientists. What an expert might sincerely dismiss as a minor risk may loom as a very large concern to people whose interests are directly affected.

The dilemma continues. How do we differentiate between real and perceived concerns? How do we balance focused, efficient identification of what really matters with involvement by those directly affected by a proposal? How do we avoid unproductive emotional confrontations?

Any answers must be credible to all those involved. In-camera sessions, token reviews, and meaningless lists of data collection and research tasks are unacceptable. The process most likely to succeed must be open, candid, thorough, and fair and, at the same time, ordered, disciplined, timely, focused, and tightly managed. In order for EIA processes to become more efficient, these criteria must be included, as a minimum requirement.

Science

Regardless of public interest and involvement in a specific EIA, more attention must be given to fulfilling information requests for scientific data. It is easier to identify and analyse biophysical effects than it is socio-economic ones, since the latter tend to depend more on value judgments. Therefore, as a first step to improving EIA, attention should be focused on developing a solid scientific basis.

Beanlands and Duinker (1983) make a significant advance in *An Eco-*

logical Framework for Environmental Impact Assessment in Canada. Their message is that the science of ecology has much to contribute to one's ability to deal creatively with environmental problems. Specifically, a comprehensive approach should involve scientific planning early in project development; project designs that treat projects as experiments from which to learn as much as possible about the ecological systems involved; continuity of design, to ensure that pre-project and post-project monitoring are consistent in both detail and focus, and a system for communicating the resulting information to those responsible for evaluating and approving future projects. Beanlands and Duinker also explore the value of ecology as the basis for improving impact assessment processes and point out that in some cases the most important ecological effects involve single species of particular economic or social importance, while in other cases ecosystem-level processes are of greatest concern.

Insufficiencies in the execution of preliminary assessments remain. There is little direct experimentation to test hypotheses, and there are few rigorous before-and-after studies to assess the accuracy of initial predictions. Such a situation suggests that much of the expenditure on EIA is not creating effective scientific knowledge. Nevertheless, most assessments, if adequately carried out and followed through can add to our knowledge and thus bring about more effective EIAS.

Unfortunately, much work done to prepare an EIA cannot be used for scientific purposes, since many of the people associated with the work, or with reviewing the completed assessment, have different goals or interests. This is especially true for the government scientist, who is mainly interested in producing good-quality scientific research and is not interested in reviewing an environmental impact statement or participating in a public hearing where he or she may be open to various kinds of cross-examination from both proponent and public. Furthermore, the scientist receives little recognition in career appraisal for this type of activity. And, in addition, individual expertise is often underused. Normally deadline requirements for preparation and presentation of an EIA do not allow sufficient time and funds for either pure or mission-oriented research. Often scientists must rely totally on information already collected in order to make reasonable assumptions and predictions about possible effects.

Scoping

It is essential to improve the capacity to identify truly important issues early in the review. Failure to identify and focus on key issues or to define the scope of assessments still remains a stumbling block, resulting in generalized and voluminous discussion of a wide range of topics, at the expense of the most important ones. The tendency to cover everything also severely limits the scientific and public credibility of the process. Efficient assessment begins with scoping, or "bounding" by determining issues to be considered and identifying their geographic, temporal, and political scope. Preliminary consultation and negotiation with chief participants can lead to agreement on an appropriate framework for analysis and review. In this way the EIA can avoid extraneous issues.

Focusing involves progressive elimination of unimportant concerns and concentration on significant ones. Subsequent scientific research and analysis can them be more mission-oriented, avoiding costly delays. Information requirements for the APP and Beaufort Sea panels were refined through public input and review and yet were still too comprehensive. (The Beaufort Sea report was more focused, mainly because of experience gained from the APP and previous EARP panel reviews.) Nevertheless, more "scoped" and "focused" EIS guidelines would have decreased costs for industry and all those engaged in reviewing the EISS and their supporting documents. However, more people might have noticed more deficiencies and called for additional work to redress these!

Bodies such as EARPS must become more efficient at scoping and focusing. If they choose to exclude certain highly sensitive issues, because enough information is already available about these, this fact must be explained to the public. The same holds true for an issue that the public wants included but has been dealt with satisfactorily in a previous assessment or through some other government mechanism or process.

Impact Management

In the face of demands for less time-consuming impact assessment, environmental managers will have to derive impact hypotheses based on short-term analyses of available facts and scientific data. Their collective professional experience and intuitive judgments will have to play a greater role. Consequently, increasing attention is being paid to mitigative measures to offset the predicted type and degree of effect and to monitoring programs to verify or refute the hypotheses.

Improving predictive ability is a major step to more effective EIA, but impact management is also important. Through experimental design and management, adjustments to project design and operation can avoid harmful results. EIA should be an interactive process involving a feedback loop from initial project design, predictions, auditing of predictions, and impact management to subsequent changes in project design. If this process is followed, future EIAS will be better equipped to deal with uncertainty, and actual effects can be better mitigated or avoided.

Although surveillance helps to improve the link between impact assessment and project implementation, it is considered a necessary evil by proponents, rather than an effective management and learning tool. Good environmental management practice makes good economic sense, and proponents should develop procedures for linking impact assessment and project implementation.

Post-Project Audit/Evaluation

No EIA, no matter how well conducted, can predict all the effects, or their relative nature, of a proposed development. Perhaps meaningful predictions are the exception rather than the rule. An audit later during project implementation allows EIA practitioners to evaluate predictions and the techniques used to make them. In addition, audits can show whether all the effects known to have occurred were identified at the assessment stage. In order for EIA to progress, practitioners must learn from experience and extrapolate lessons to other situations. Post-project audit accomplishes this goal. Without such analysis, the rationale for EIA can be called into question.

Effective audit of a project EIA requires that the evaluation itself be kept in mind at the beginning of the assessment process – during baseline studies, formulation of predictions, writing of the EIAs, design and implementation of monitoring schemes, and design of mitigation measures. Neglect of this approach has proved some projects' downfall.

The conduct and completion of post-project evaluation presuppose certain enabling conditions. There must be regular measurement of both biophysical and socio-economic effects, to determine how well environmental resources, community values, and land use opportunities have been safeguarded. This kind of information is absolutely essential to establish cause-and-effect hypotheses. Government-funded research on long-term baseline monitoring and development of a working knowledge of processes are further prerequisites for improved predictive capability. More than anything else, lack of post-project monitoring has held in check the advance of impact assessment.

CONCLUSION

In the future, the primary need for sophisticated marine transport will depend directly on the fortunes of the mineral resource companies, particularly in the offshore, as both Canada and the United States look to the Arctic as one of the few remaining resource-rich frontiers. Economics will be the principal factor dictating the rate of expansion of arctic marine transportation. The forecast for arctic shipping over the next two decades is for dramatic increases in vessel size, numbers, frequency of transit, and length of shipping season. This increase will result from development of offshore oil and gas, mining operations, and community resupply activities, in response to growth in population, income, and consumer consumption.

In early 1985, the federal government and the government of Northwest Territories announced that they would allow the first crude oil in the Canadian Arctic to be produced and shipped through the Northwest Passage. Panarctic Oils Ltd was given approval to proceed with its Bent Horn oil-field demonstration project, with two million barrels of oil being shipped south to a Montreal refinery over five years. The oil is being shipped by an ice-reinforced tanker during the short summer season only.

The project has been designed for a tanker to make the voyage once each year under optimum summertime conditions, with the possibility later of a second summer voyage. Throughout the final 100 km leading to the well-site, where the ice is at its thickest, a Canadian Coast Guard ice-breaker will, if necessary, accompany the tanker, to provide additional safety. In granting approval, both governments stressed effective monitoring, which would include arctic native input and co-operation, to observe any adverse environmental effects.

The Bent Horn Project is a "foot in the door" for future arctic shipping interests and provides the first small-scale opportunity to test arctic shipping by tanker. If not carefully controlled and monitored, however, the project will harm future shipping interests in the region.

Scientists from all levels of government, academe and industry can learn from this project. In this "working laboratory," many environmental issues highlighted by the APP and Beaufort Sea panels can be evaluated. Previously, scientists were aware of problems in arctic marine transport, but could not test their solutions experimentally.

In conclusion, EARP has contributed significantly to the scientific and decision-making regime of arctic shipping, and vice versa. Although many concerns remain, the process has evolved considerably, to become perhaps one of the more credible strategies in international circles. However, improvements discussed in this chapter, such as experimental design and more effective scoping, can and should be made as soon as possible. Senior decision-makers should recognize that information obtained through science and EIA can greatly assist the decision-making process.

PART THREE

The Big Picture

CHAPTER NINE

Hydrocarbon Transport and Risk Assessment

Ray Lemberg

The analysis of risks to the environment or to human life or health is becoming increasingly common. These risks may be produced by various courses of action or may be likely to be produced in the future.

Everybody performs risk analysis every day. Before crossing the street, a person hesitates until cars are judged to be at a safe distance. In this chapter, "risk analysis" refers to a formal process involving quantitative assessment, valuation, and management of risks.

BASIC CONCEPTS

Risk

The work "risk" is often used as if everyone knows its meaning, but for the purpose of quantitative risk analysis the word should be defined. There are as many definitions of risk as there are authors on risk analysis. Some writers avoid the word altogether. One dictionary defines risk as: "the chance of injury, damage, or loss" (Webster 1977). Risk implies chance or uncertainty and is often unavoidable. In mathematics, a chance is called a probability. For example, the probability that a randomly chosen person will die from an automobile accident is about 0.02 per cent per annum. The risk of death from an auto accident has three basic components: an event (the automobile accident); a deleterious consequence (death); and a probability that the event will take place (0.02 per cent per annum per capita). This combination (triplet) of event, consequences that may ensue, and probability that the event takes place and the consequences ensue is the definition of risk used in this chapter.

The triplet can be condensed into an expectation of risk. For example, if a million people suffer a risk of one in a million from some specific automotive problem (say, bad brakes causing a fatal accident), one person is expected to die each year.

Following the example above, the question may arise as to whether a one-in-a-million risk is acceptable. Often it is, when applied to a total population. However, if one could identify which of the million people is to die, the risk would seem much less acceptable, for then it is no longer a risk. It becomes a certainty.

In risk, one does not generally know how to identify the recipient of the undesirable consequences of a risk. In the example above, who will die from the risk of bad brakes, or which automobile's brakes will fail? If one did know, one could usually prevent the doomed person's fate. Decisions about risk abatement therefore are made difficult by the fact that the risk to the whole population must be reduced in order to save the one person in a million.

Hazard

A hazard is defined as something that is inherently dangerous to life; it does not require harm to occur. For example, a tree falling in a forest is a hazard even if no living thing happens to be under it when it topples.

What is the difference and why distinguish between risk and hazard? Hazards are often easier to assess than risks and generally result from technological or industrial processes. For example, a coal-fired electricity-generating plant will release a certain volume of sulphur dioxide into the air per year. These emissions may be taken as a measure of the sulphur dioxide hazard, whose value may be assessed fairly accurately from engineering data and measurements of stack emissions.

Much more information is needed to determine the risk to human health presented by the sulphur dioxide hazard. One would need to know the wind patterns and population numbers around the plant, the chemistry and sulphur compounds in the atmosphere, the way in which outdoor pollution levels correlate with indoor levels, and the relationship between exposure to pollution and health effects.

Assessing the risk to the environment uses the information about the hazard and requires further understanding of the interactions of the ecological components involved, an understanding that may be rudimentary at best.

In general, one knows more about hazards than about risks, which has led to hazards being used as surrogates for risks. Regulators tend to deal with hazards, rather than risks. For example, standards for sulphur dioxide emissions are expressed not in terms of the number of deaths or watery eyes allowed but in terms of kilograms of sulphur emissions allowed per ton of coal burned.

Regulation of hazards helps alleviate to a large extent the burden of enforcement. Enforcement becomes a matter of physical measurements, rather than epidemiology. But hazard regulations are likely to have very different effects from regulations to reduce risks. For example, risk-reducing regulations may allow higher pollution levels in unpopulated areas, yet apply strict controls in densely populated cities.

Environmental Risk Assessment

The study of environmental risks is open to great analytical subjectivity. Even in the risk measurement stage, the question of the criterion to define a risk is far more obscure than in human health risks, because so little is known about ecosystems and their capacity for coping with stress.

Ecosystem stress should be an objective and measurable characteristic of the physical impact or disturbance, namely, the hazard. Yet, even at this stage, precision may not be possible, because the potential hazard cannot be specifically identified.

"Stability" and "threshold" are two concepts used by ecologists in the attempt to measure ecosystem disturbances. Stability is a measure of the strength of an ecosystem under stress. Current ecological doctrine holds that established ecosystems are resilient to small stresses. Even ecosystems described as fragile are fragile only to stresses that are not present in the long term. Thus arctic tundra is fragile in terms of some types of human interference but very resilient to variations in the severity of winter weather.

The threshold is that point at which a particular stress will produce irreversible changes in an ecosystem. Stress below a certain level will make no significant long-term differences. Stress beyond that level will cause irreversible damage and change.

Unfortunately, the concept of threshold presents difficulties in risk analysis. Ecosystems are very non-linear in their response to stress. Increasing the stress by a small amount can change the ecological effect by an order of magnitude or more. Predicting the ecological effects of stress requires accurate knowledge of threshold levels, which is available for only a few stresses on a few species in a few ecosystems.

Stresses can be considered minimal if they are within the natural variations that any species must be equipped to survive. Any wild species must survive changes caused by weather, fires, disease, and the like, which can kill several percent of the population in a given year. This principle can be very useful in screening out minimal hazards and in some cases may be all that is needed to evaluate energy-related environmental risks.

The ideal of environmental risk analysis would consist of a dynamically evolving population model for an ecosystem. This model would allow the stresses of interest to be entered as parameters and would give the ecosystem's response both while the stress is being applied and after it is removed.

At present, ecological science is not capable of constructing such models in most cases. In some circumstances, models can be used successfully if the population dynamics equations are particularly simple. For example, it is possible to use the known life-cycle of some food fish to model the response of the fish population to various harvesting levels by commercial fishers. The risk is the chance that a fish will be caught by the boats fishing an area, and this is predictable from information about the fishing fleet. Since the risk does not affect those fish not actually caught, the population dynamics equations are simple.

In most environmental risks, the elements are not particularly simple. For example, when considering a chemical contaminant, the mortality of individuals in a bird population may be much less important than changes in breeding efficiency. DDT, for example, does not kill many birds, but it could kill off a species. Usually, not enough is known about such questions to deal with them at the risk level, and proxies based on the hazards involved are therefore often used instead.

Difficulties in Risk Assessment

Hazards are assessed by analysing the technology that produces the potential hazard. Technology assessment is probably the least difficult of all environmental risk analysis activities, since more is usually known about a particular project technology than a disturbed environment. Many sophisticated analytical tools exist to determine the probabilities and severities of the various hazards to an environment.

The assessment of the environment effect of hazards is not only more complex but also suffers from a lack of scientific knowledge and data. Whereas assessment of a technology can be accomplished fairly quickly, much more time is usually required to obtain the necessary environmental knowledge and data – often more than the period allotted for obtaining a decision about a project's environmental impact.

Perhaps the most difficult step for decision-makers is to apply value judgments to the consequences of environmental risks. It is a subjective exercise, influenced by the natural biases of the evaluators. It is also usually performed in the absence of clear precedents or guidelines for determining acceptability.

Another fundamental difficulty in the assessment process involves establishing boundaries both in time and in space (Beanlands and Duinker 1983). Boundaries for project elements (the technology) are relatively simple to establish. Physical boundaries are usually well defined, and economic considerations determine the time-span of the construction and operations phases of a project.

Environmental boundaries, in contrast, are much more difficult to establish. The setting of such boundaries represents a compromise among the constraints imposed by political, social, and economic realities (administrative boundaries); the spatial and temporal extent of the project (project boundaries); the time and space scales over which natural systems operate (ecological boundaries); and the limited state-of-the-art knowledge in predicting or measuring ecological changes (technical boundaries).

Some of these categories are under the control of investigators, while others are relatively fixed.

RISK ASSESSMENT OF SPILLS

Discussion of risk analysis in this chapter focuses on episodic spills of oil or liquefied natural gas (LNG). More comprehensive discussion would cover chronic and episodic hazards, including spills, noise, chemical discharges, and disturbances to ice and habitats caused by the passage of ships. Spills of oil or LNG resulting from accidents are classified as episodic hazards: low probability of occurrence with potentially large spill sizes. These large spill sizes are cause for environmental concern, and there is interest in how such hazards are assessed.

Spill Hazard Assessment

Assessment of the risk of episodic oil or LNG spills in the Arctic falls within the definition of hazard assessment as discussed above. The objective of an analysis is to determine the probability of occurrence for spills of various sizes. Whether dealing with oil or LNG, the spill risk assessment consists of determining the probability of occurrence of an accident that might lead to a spill, the probability of occurrence of a spill given that an accident has occurred, and the probability distribution of the spill size given that a spill has occurred.

Many sophisticated methodological tools are available for determining these three probabilities. The fault tree analysis method may be employed to determine the probability of an accident's occurrence. Assuming an accident has happened, engineering analyses can determine the degree of structural damage to a ship. The location and severity of damage, compartmentalization of the cargo, and probable response times to shut off the release of the cargo are used to determine the probabilities of various amounts of spillage.

While the tools used are sophisticated, the results obtained must be viewed with caution. There is an "Achilles' heel" to the analytical process: we have no experience for arctic marine transport of LNG and only limited "pilot experience" with arctic oil transport. There are considerable world-wide data about shipping accidents and equipment failure, but these have to be adjusted for arctic marine transport, and these adjustments naturally involve judgmental decisions and possible biases of those performing the analysis.

For example, the design of arctic tankers is different from that of the average tanker. The proposed arctic tankers are designed to be structurally stronger, have ice-breaking capability, the latest navigational aids, strengthened cargo containment compartments, redundant propulsion, and operational systems, and their crews will receive more extensive training.

While such design features may reduce the chances of a spill, the amount of reduction is usually a matter of expert opinion. Expert opinions, however, do not always coincide, but they do point out inherent weaknesses in assessments and thereby invite debate. At the very worst, the assessments will not provide an absolute answer but will instead define the limits of the problem. These limits will depend on the degree to which expert opinions may differ and the degree to which they can be supported.

A spill risk assessment is unlikely to produce unassailable conclusions. But the upper and lower limits of the problem can be established, which will encompass the range of expert opinion. With boundaries established, it should be possible to proceed to the next step, determining the environmental effects which may result.

Spill Assessments to Date in the Arctic

Two proposed hydrocarbon transport projects in the Arctic have been subjected to the Environmental Assessment Review Process (EARP) to date. The Arctic Pilot Project (APP) proposed production and liquefaction of 6.4 million cu m of natural gas per day from the Drake Point field on Melville Island and its shipment to eastern Canadian markets in ice-breaking tankers. The shipping component of this project would have been the first attempt to operate year-round in the High Arctic and the first LNG transport requiring break-up of heavy ice during normal operations (FEARO 1980).

The proposal for Beaufort Sea hydrocarbon production and transport described a range of options and scenarios for oil and gas production and transport from the Beaufort Sea. These options ranged from oil production of about 15,000 cu m per day to the "highest technically achievable" level, of about 200,000 cu m per day. Oil-carrying tankers were proposed as one option for transporting oil from the Arctic (FEARO 1984).

The transport components of these two projects contained many similarities in tanker design and operation. The tankers would be massively stronger than conventional tankers and two to three times stronger than required by existing legislation. Their design includes many safety features not found on conventional tankers, including a double hull, which would have a double bottom to reduce the danger of a spill in the event of an accident, and cargo holds within the inner hull, so that no cargo would be carried adjacent to the outer hull. These features would reduce, but not totally eliminate, the risk of a spill in a moving accident such as a collision or grounding.

LNG *Spill Risk*: The history of LNG carrier operations suggests that the risk of an LNG spill is small. Since 1964, over 3,100 voyages have been completed by more than twenty-five different LNG carriers. Shipments have been made from six liquefaction sites to receiving terminals in eight different harbours along routes dispersed over international shipping lanes. In the more than 120 ship-years of day-to-day operations, there have been no accidents endangering the public, no loss of ships, nor any fatalities to crews because of LNG spills. This excellent record can be attributed to the special design and equipment of the carriers, careful crew selection and training, and the special safety procedures used for navigation and ship manoeuvring.

About a dozen incidents of LNG carrier problems have been reported, mostly equipment failures (readily corrected) and design deficiencies, of the kind usually encountered in most new operations.

From the world-wide record of LNG carrier operations, the annual frequency of an LNG spill is less than 1 in 120 per ship year. As proposed, the APP's LNG carriers are structured to meet or exceed the design, operation, and safety criteria of LNG carriers currently in service worldwide. The additional safety features and increased strength may reduce the annual frequency of an accident by as much as a factor of 100, as estimated by proponents. For the proposed two-carrier operation, the annual accident probability would be smaller than 1 in 100, and maybe as low as 1 in 10,000 (APP 1979).

While the possibility of serious damage during LNG carrier operations is small, it cannot be dismissed altogether. In the event of damage, the worst case that could reasonably be expected would spill LNG from two of the six tanks on board. It has been estimated that combustion of the spilled LNG would produce intense heat but with effects limited within an 11–km radius of the ship. This range was based on predicted diffusion and dilution rates and the fact that LNG is combustible only in 5–15-per cent mixtures with air. While this prediction is based on limited evidence, the APP Panel believed it acceptable for assessment of the spill risk.

Oil-Spill Risk: The proponents of the Beaufort Sea oil transport proposal estimated the risks of oil-spills from arctic tankers (Bercha and Associates Ltd. et al 1983). They adjusted world tanker statistics to reflect the more severe environmental conditions expected in the Arctic and to allow for the safety features built into the proposed arctic tanker. They concluded that their proposed design would be many times safer than a conventional tanker operating in other waters and as much as one hundred times less likely to have an accident.

The proponents indicated that tanker accidents involving collisions with another vessel, ramming an iceberg, or grounding would be unlikely to cause loss of a whole ship. In these situations, loss of up to three cargo tanks out of sixteen could occur. An accident involving loss of an entire cargo would be quite unlikely but could conceivably occur because of fire, explosion, or structural failure.

Calculation of the increased safety factor was a controversial subject among experts (FEARO 1984). Are increases in safety caused by design and operational improvements cumulative? Furthermore, there is considerable engineering judgment involved in evaluating the effect of each design feature. For the proposed two-tanker operation, the proponents claimed an annual oil-spill frequency of less than 3 in 10,000 including the 100-times reduction factor. If all the reduction factor is removed, the annual frequency increases to less than 3 in 100 per year.

Proponents cited research indicating that 75 per cent of tanker accidents involved human error. In some cases, extreme weather conditions, such as high seas or fog, contributed. Very few tanker accidents have been caused solely by human error. Human errors have been a result of inadequate training, flouting of the law, breakdowns in communication or authority in vessel operation, honest mistakes, drug- or alcohol-induced accidents, and errors in equipment design. Proponents intended to control these sources of error through careful recruiting, training, and supervision and use of fail-safe designs.

Proponents described their suggested safety programs to reduce the human error factor in accidents. Their safety record in Beaufort Sea operations to date has been excellent. If this trend were to continue when the proposed tankers are operating, methods to reduce human error would become the most direct means of avoiding a major oil-spill.

There was considerable debate and difference of opinion about the size of oil-spill resulting from an accident to a tanker. This vessel would be designed to operate under arctic conditions year-round with a cargo capacity of 240,000 cu m, contained in sixteen cargo tanks. Proponents estimated an average oil-spill of about 12,000 cu m. After considerable discussion with proponents and other experts, it was concluded that the maximum credible spill resulting from a collision, ramming, or grounding accident would not exceed 41,000 cu m. For an accident caused by structural failure, explosion, or fire, the maximum credible spill size was estimated as 220,000 cu m. By comparison, world-wide statistics revealed that the biggest recorded oil-spill was about 240,000 cu m.

Numerous vessel features contributed to acceptance of the 41,000 cu m maximum. The arctic tanker would have separate oil cargo and ballast tanks. In the event of damage to a cargo tank, the oil could be transferred to an undamaged ballast tank. The ship would also be equipped with dual inert gas systems, which would be used to flood all open cargo spaces with inert gas to reduce the chance of explosion caused by ignition of volatile substances in the ship's hold. The risk of explosion would be further reduced by the use of deep well pumps, individually sited on the deck above each cargo tank, rather than a single pumping station within the ship.

ENVIRONMENTAL EFFECTS OF TANKER ACCIDENTS

The spill hazard assessment determines the probability distribution of spill size from a particular tanker. Environmental impact assessment (EIA) starts with this information and, for a particular spill size at a particular origin, determines the exposure of environmental components (recipients) to the spilled oil or LNG and then determines the resulting consequence to recipients.

Exposure determination is a complex and difficult exercise. Recipients have to be identified. A recipient may be an organism, a species, a population, or an ecosystem. Baseline environmental data ought to be available to describe their locations, possible migratory patterns, abundance, and interdependencies. One must next identify pathways by which the various recipients may be exposed to the spilled oil or LNG. Spill dynamics (movement of the spilled quantities over space and time, plus any changes in their chemical composition) must be predicted.

Each pathway may, of course, present a different concentration of spilled substance to recipients. The final step is to determine the consequence to the recipient of exposure to particular concentrations of the spilled material through the various pathways.

As previous EARPs have shown, EIA results in qualitative, descriptive assessments employing phrases such as "major," "moderate," "minor," and "negligible" effects.

Difficulties in Arctic EIA

There are many reasons why EIAS in the Arctic cannot be more quantitative or precise. Examination of three major difficulties in the assessment process reveals why.

First, very few baseline environmental data exist for the Arctic. Data-gathering has been motivated primarily by proposed projects, and there has been little research not sponsored by proponents. If projects remain the primary motivator for obtaining baseline data, improvement may be anticipated. The relatively short time-frame over which EARPs are conducted provides insufficient time for environmental research. Research programs spanning a decade or more are required to obtain data with sufficient statistical validity about trends and natural variations in the abundance of environmental components. Otherwise, potential disturbances caused by a project may very well be attributed to natural variation, and it will be virtually impossible to separate cause and effect.

Second, predicting pathways is difficult. The trajectory models used to predict the movement of oil or LNG from the source of a spill are at best approximations. They use large-scale meteorological data to predict much smaller-scale movements of oil or LNG. Local meteorological conditions (wind, waves, currents, ice movements, temperature) can differ considerably from large-scale meteorological conditions. Furthermore, tanker routes from the Arctic are long, and accidents could occur anywhere along the route. Even if local conditions were known all along the route, prediction of spill trajectories, concentrations, and chemical changes for each possible spill location would be a formidable computational task.

Third, a recipient at a particular location may be exposed to different concentrations and compositions of hydrocarbons which may reach it along numerous pathways. Total exposure of the recipient must include all possible pathways – yet another formidable computational task.

Arctic EIAs to Date

The federal EARP has provided EIAs of two proposals involving hydrocarbon transport in the Arctic, through hearings conducted by two EARP
panels: for the Beaufort Sea (FEARO 1984) and for the Arctic Pilot Project
(APP) (FEARO 1980). These two transport proposals contained some common issues relating to the effects of ice-breaking on seals, the passage
of hunters and migratory animals across the ice, and the effect of underwater noise on marine mammals. These hazards are not the subject
of detailed consideration in this chapter, although they provide some
of the same problems of assessment as the effects of oil or LNG spills.

The methods used by proponents, intervenors, and others to assess
the potential environmental impact of the two proposals had limitations.
Environmental impact statement (EIS) documentation provided a good
overview of the problems that can be caused by oil or LNG transport.
Prediction and assessment of effects are made difficult by data deficiencies concerning animal species; lack of methods for assessing potential cumulative and synergistic effects of development; logistical
problems in gathering data in arctic environments; the conceptual nature
of proposals; and limited understanding of the level and significance of
effects.

The review of both projects provides a good overview of data deficiencies and can also help assign priorities for future research and
monitoring. However, assessment of biophysical effects cannot be considered complete. The EISS represent a significant achievement, but ought
to be viewed as a starting point.

The EIS for the Beaufort Sea proposal, being the most recent, is
perhaps the most comprehensive to date (Dome, Gulf, and Esso 1982).
The matrix approach used is a simple and effective way to summarize
most of the perceived environmental consequences of development. It
serves to identify broad areas of concern, but many of its conclusions
are based on professional judgment, based often on experience from
other areas.

Proponents' method was criticized by several review participants as
containing inaccuracies and inconsistencies and as not being sufficiently
specific or comprehensive. Very broad impact categories, such as "major," "moderate," "minor," and "negligible," were criticized as im-
precise and often misapplied. The matrix technique was called inadequate
and unable to model cumulative and synergistic effects. Many criticisms
centred on problems generic to EIA. Although the method of proponents' EIS had many critics, few presented a practical alternative.

The review process identified the need for effective and timely environmental research and monitoring, but how do we determine the

significance of environmental effects? Without criteria of "significance,"
it is difficult to monitor projects and to determine if an actual effect is
adverse or diverges markedly from an anticipated expected effect.

Impact of LNG Spills: The APP EARP Panel concluded that birds might be
at risk as a result of an LNG spill:

The principal risk to birds would be to Murre colonies at Prince Leopold Island,
Cape Hay and along the west Greenland coast, where these birds are locally
concentrated on the water close to the colonies, and to a lesser extent to concentrations of feeding birds along the coasts and ice edge. In late summer the
Greenland coast north of Disco Island is used as a moulting area by Common
and King Eiders feeding on northern Baffin Bay and much of the Canadian
High Arctic. In an LNG accident the birds in such concentrations might suffer
significant mortality from freezing if the LNG did not catch fire and from burning
if it did. The Panel believes that the carrier route would be too far offshore for
an accident to endanger such concentrations, though flocks of birds feeding at
the ice edge would still be vulnerable. (FEARO 1980:72)

Presumably humans and other mammals within the danger zone could
suffer the same fate.

Impact of Oil-Spills: Proponents of the oil transport proposal stated that
their normal offshore operations would generally result in only negligible to minor effects on offshore species. Any long-term effects would
be localized and not of regional significance. Detailed discussions took
place about several important species, and the findings of the panel are
summarized below.

  (1) *Polar bears*. Transport of oil can harm polar bears. If a polar bear
is oiled, it would lose its thermal insulation and would die (FEARO 1984:
section 6.7.2).

  (2) *Seals*. "A major oil spill would seriously disrupt local populations
of seals, particularly if the oil accumulated below sea-ice and blocked
the animals' access to dens or breathing holes. Even in the worst cases,
significant impacts to local populations would probably be offset by
the resilience and wide distribution of the species. The more subtle
chronic effect on food sources resulting from such occurrences could
also continue to impact seal populations. The Panel concluded it is not
possible to reach definitive conclusions on the potential long term impacts on these species" (FEARO 1984: section 6.7.3).

  (3) *Whales and Walruses*. The effects of a large release of oil on whales
are not understood. There are still some significant gaps in the knowl-

edge about the biology and ecology of whales. The panel could not draw conclusions about effects on walruses, because no specific data were presented. It concluded that continuing research would be required, because walruses might be affected by tanker traffic (FEARO 1984: section 6.7.4).

(4) *Fish.* "The Department of Fisheries and Oceans informed the Panel that before the effects of any disturbance on fish can be assessed, it is essential to have broadly based knowledge of the system likely to be affected. Such knowledge of Arctic fish is not yet available and logistic difficulties make the required ecological data difficult to obtain. For instance, Arctic cod, although not directly harvested, is the fish species of greatest significance in the Arctic marine food chain. Local disruptions to such populations could cause considerable impacts on birds, seals, and other animals that depend on Arctic cod for food" (FEARO 1984: section 6.7.5).

(5) *Marine Birds.* There is little information on the seasonal distribution and ecology of many species of marine birds. The Canadian Wildlife Service informed the panel: "Significant portions of the areas' breeding populations gather in small areas of open water, particularly during the spring migration in most years, and throughout the breeding season in heavy ice years, where spilled oil is likely to collect. Any bird contaminated by oil in such conditions will most likely die. In addition, survival of some populations could be further jeopardized by changes in food availability caused by disruption or contamination of feeding areas associated with stable ice edges. Further, there is no evidence to suggest that available oil spill countermeasures can mitigate these adverse effects" (FEARO 1984: section 6.7.6).

(6) *Marine Organisms.* Oil-spills and chronic discharge of pollutants could significantly harm localized populations of marine organisms: "The Department of Fisheries and Oceans (DFO) stated that, in the case of a massive release of oil, there could be significant effects on the subtidal flora and fauna and the under-ice biota. As these are probably the two most productive systems in the Arctic marine environment, DFO felt that more basic research would be of value to help define long term impacts" (FEARO 1984: section 6.7.7).

Conclusions of the Panel

The Beaufort Sea EARP Panel did not recommend oil transport by tanker. Instead, it recommended that: "The Government of Canada approve the use of oil tankers to transport Beaufort Sea oil only if: (a) A comprehensive government research and preparation stage is completed by

governments and industry, and (b) A two-tanker stage using Class 10
oil carrying tankers demonstrates that environmental and socio-economic
effects are within acceptable limits" (FEARO 1984:104).

Unfortunately, the panel did not specify "acceptable limits." The
inadequacy of existing information on population and behavioural ecology of major arctic marine species hampers careful analyses of undisturbed environments and impact assessment. In many areas of concern,
the environment is already being altered by human activities, so that
the opportunity to obtain baseline biological data is fast slipping away.

CONTINGENCY MEASURES

As the term *contingency* implies, these measures are to be applied only
after a spill has occurred. Preventive measures are incorporated into
tanker design and operations. Contingency measures generally include
four activities: stopping release of hydrocarbons from a tanker; containing the spread of the spill; cleaning up the spill; and repairing damage
to shorelines.

For LNG Spills

Other than shutting off the flow of released LNG, the only effective
contingency measure is evacuation of people within an 11-km radius
of the ship, who might be exposed to fire (FEARO 1980).

Spilled LNG vaporizes, and the gas will spread as it mixes with air.
It will drift downwind as a ground-hugging cloud, made visible by the
condensation and freezing of atmospheric water vapour, and it will rise
as it warms. Part of the initial cloud may spread upwind some distance
due to gravitation, but eventually, as it warms, all of it will be swept
downwind. If the cloud encounters no ignition source while the vapour
mixture is within the combustible range (5–15-per-cent mixture with
air), it will dissipate harmlessly into the atmosphere. However, if the
mixture is ignited, it will either burn locally or spread some distance
back or all the way back to the spill site and form a pool of fire. Burning
will continue until the fuel supply is exhausted or the flame is extin-
guished. The LNG vapour cloud could explode, producing a potentially
destructive pressure wave.

Calculations can be made to determine the maximum distance from
the site of a spill in which a potentially dangerous environment could
exist. The greatest danger in this environment is posed by drifting
vapour clouds. For the APP, the safety distance for a spill of two LNG
cargo tanks (56,000 cu m) was predicted to be about 11 km. This estimate considered a number of factors: volume and rate of LNG spillage;

the surface on which it is spilled; degree and geometry of liquid confinement; and location and timing of ignition.

For Oil-spills

Before discussing post-spill measures, it is useful to understand the general behaviour of spilled oil in the arctic environment. The behaviour and characteristics of spilled crude oil depend on its chemical and physical properties and on the physical environment into which it is spilled. Oil discharged on open water undergoes several processes that affect its fate and behaviour.

Spilled oil will spread quickly, resulting in a thin slick that covers a large area of the ocean's surface. The slick will also drift with wind-induced surface currents and will be transported by the residual water movements in an area around the source of the spill. Turbulence in the near-surface water will break up the oil slick into patches once it is thin enough. These patches will generally be surrounded by thin sheens, which appear to be fed by the patches.

Along with spreading and movement, the slick will weather, a process that will alter its composition. Weathering processes include: the formation of emulsion (a relatively stable mixture of oil and water), which increases the slick's viscosity and its volume on the water surface; dispersion, which reduces the volume of oil on the water surface by mixing some of it into the water column; evaporation, which results in the rapid loss of the lighter fractions of the oil into the atmosphere; sedimentation, the sinking of oil droplets by their attachment to suspended sediment; dissolution, which leaches out the water-soluble oil fractions; and various oxidation and bio-degradation processes, which slowly change the chemical make-up of the oil and ultimately remove it from the sea.

The long-term result of oil left on water is "tar balls" mainly the very heavy ends of the oil. The heavy ends can have a density equal to or greater than water and may be suspended throughout the water column or settle onto bottom sediments.

Clean-up measures depend on whether the oil-spill has occurred in open water or in ice.

Clean-up in Ice: Oil discharged beneath ice will ultimately appear in melt pools during spring break-up. The Beaufort Sea EIS describes *in situ* combustion techniques to be applied to oil in melt pools and other oil trapped or contained by ice. This appears to be the only oil removal technique in ice-covered waters.

In situ burning is a one-step removal process and eliminates the need for containment, mechanical recovery, transfer, concentration, and disposal of oil. The success of burning depends on air-deployable igniters. During winter, the igniters could be used to burn oil contained in melt pools between ice floes or on ice.

It appears that the *in situ* combustion of oil trapped in ice-covered waters ought to be substantially completed before open-water conditions occur; otherwise, one would have to revert to open-water clean-up methods. The oil-contaminated ice must be located and tracked throughout the winter so that the oil appearing in melt pools can be ignited.

The EIS proposes to use positioning buoys that can be tracked by satellite to mark spills. Presumably, the igniters would be dropped during the spring into melt pools containing the oil.

This clean-up technique is advanced by studies reported in the EIS. However, its success will depend ultimately on the availability of radio-tracking buoys, the associated satellite communications system, a size-able fleet of helicopters to deploy buoys and drop igniters, and follow-up monitoring to ensure that the spilled oil has been removed.

The EIS does not indicate the exact extent of these resources to be available close to the Beaufort Sea area and the Northwest Passage. Given the limited range of helicopters, one can envisage shore bases distributed through the Arctic.

The tanker transport route through the Northwest Passage covers the largest and longest region to be protected. If one operator in the Beaufort elects tanker transport and another chooses a small-diameter pipeline to Norman Wells, could the tanker proponent economically justify deploying the necessary clean-up resources to protect the Northwest Passage?

Clean-up in Open Water: Clean-up in open water consists of five steps: containment of the oil-spill, removal of the contained oil, transfer of the removed oil to a processing facility, separation of water from the recovered oil, and disposal of recovered material, including oil and oil debris.

The spread of an oil slick may be contained by deploying floating booms. According to figures in the EIS, approximately 10,000 m of various sizes of booms are stockpiled in Northwest Territories. The EIS does not indicate what size of spill can be contained with the available booms, or the additional stockpiles required to handle a large spill volume, such as that from a well blow-out. Booms are not a fail-safe method, because, depending on the booms' size, excessive current velocities, winds, and waves can splash oil over the top of the devices.

Physical removal of oil from the water surface may be accomplished in various ways, but all of these depend on the presence of a response barge, which contains the removal system, as well as other equipment for water separation and storage of removed materials. The response barge mentioned in the EIS can remove 400 cu m of oil per day. The EIS does not state how many barges may be required to respond to a large oil-spill, such as a well blow-out, or how long it may take to clean up such a spill. Containment of a large spill would require many booms, and the number of response barges available will determine the time it takes to clean up the spill. A secondary effort may also be required to contain and remove the oil that escapes over the booms.

The impact of an oil-spill on the environment will depend on the proponents' ability to contain a spill and clean it up as quickly as possible. The EIS does not indicate the time required to clean up a spill or the degree to which various levels of containment and clean-up may reduce damage caused by hypothetical oil-spills (volume 6). To increase the response also increases the cost, and disputes may arise as to the appropriate level of effort.

Protecting and Restoring Shorelines: The EIS discusses protection priorities for shorelines. Primary responses are to be allocated to environmentally sensitive areas, and appropriate plans of response and methods of shoreline restoration will be selected. The EIS proposes deployment of booms to protect sensitive shoreline areas but does not discuss how much boom must be stockpiled. Appropriate response depends on prepared plans and classification of shorelines' sensitivity. If such data are collected, response options and plans can be prepared and approved by the appropriate environmental authorities, and selection of a response option ought not to be a problem.

While the EIS discusses shoreline restoration, it does not consider the resources (equipment, manpower) required. The EIS estimates that in some spills several hundred kilometres of shoreline may be exposed and may require clean-up – a formidable task, requiring large amounts of equipment, manpower, and logistical support. How much time will it take to clean and restore so much shoreline? And who will pay?

R&D Clean-up Technology: At the Beaufort Sea hearings, proponents reviewed available techniques for cleaning up oil-spills in the Arctic and estimated their probable success in different locations and weather conditions. Companies operating in the Beaufort Sea have invested capital in buying and maintaining oil-spill equipment in the region. They have also invested heavily in technology research and development (R&D)

for oil-spill counter-measures in arctic waters. Yet available equipment is suited only for small spills. The technology to handle larger oil-spills is still in the R&D stage.

Even the best technology will probably not totally clean up oil-spills. High waves, strong currents, or certain types of ice could defeat even the best-prepared and most conscientious operator in the Arctic. Where such phenomena are present, the spilled oil would be left to evaporate, disperse in the water column, or spread over shorelines.

The costs of cleaning up spilled oil in the Arctic are very high (as much as several thousands of dollars per cubic metre). It may be more cost-effective to prevent spills, through careful monitoring of equipment, procedures, and personnel. Public pressure, however, will probably require the ability to clean up oil-spills rapidly, to levels that would leave the environment safe for arctic wildlife.

The response capability could be expressed in terms of the minimum oil recovery rate (in cubic metres per day) that can be mobilized and maintained within a specified time before help needs to be obtained from other regions. Quantifying such response would require consideration of spill location, available technology, costs, volumes of oil transported, and prevailing environmental conditions.

MONITORING

The definition of monitoring is intended to exclude the surveillance and inspection undertaken by regulatory agencies: "The term monitoring refers to repetitive measurement of specific ecological phenomena to document change primarily for the purpose of (i) testing impact hypotheses and predictions and (ii) testing mitigative measures" (Beanlands and Duinker 1983). The most effective way to determine and assess the effects of development is to allow phased development, with extensive environmental monitoring. However, it would be prohibitively expensive to monitor every potentially affected species for the entire life of a project.

Monitoring is usually discussed in connection with a specific project. Without a project, environmental monitoring would lack focus, and efforts and resources would have to be expanded over large areas to obtain baseline environmental information. Is it reasonable to establish monitoring programs along all possible transport routes through the Arctic? Could the effort be sustained from year to year? Continuation is important, since interruption or reduction of monitoring would prejudice its scientific validity and the credibility of the data. However, recent experience shows that tight fiscal policy renders unfocused environmental monitoring as of low priority.

IMPROVEMENTS TO EARP

Based on the experience at the Beaufort Sea EARP, improvements may be made in handling scientific disputes and in defining "acceptable" risk or hazard.

Scientific Disputes

The Beaufort Sea public sessions showed how little factual information is available about arctic ecosystems. Consequently, discussions became based on subjective opinions, and disagreements were evident.

In one instance, disagreement over the probabilities and magnitudes of potential oil-spills threatened to throw the public session in Inuvik off schedule. Much of the discussion focused on methodological issues, such as the merits of various statistical techniques and databases. The panel took a novel approach to the controversy:

Dr. Ray Lemberg, a Technical Specialist, at the request of the Panel ... consulted with the Proponents and their experts, as well as with the Department of the Environment and other Technical Specialists, to provide a summary of the major issues and disagreements remaining among those who had participated in preparing or advising on the preparation of the Proponents' report on Oil Spill Risk Assessment. Under the present state of knowledge about risk factors, the estimates given [to the panel] were considered by those consulted by Dr. Lemberg to be reasonable estimates of the most extreme spill sizes which could be expected in the North as a result of the Proponents' proposals. (FEARO 1984: section 4.1.1)

Technical specialists "are retained by Panels to provide impartial expert opinions and scientific facts on certain issues, and to raise other concerns which may be overlooked. Technical Specialists are available to all review participants, including the Proponents." The panel suggested that "future EARP panels continue to make use of Technical Specialists to help clarify and resolve contentious issues at the public sessions. In addition, the Panel concludes that future EARP panels should acquire Technical Specialists early in the review process, preferably at the time the EIS guidelines are being prepared" (FEARO 1984: section 2.10.).

Acceptable Risk or Hazard

Typically, the risks of most concern to the public have serious consequences but a very low probability of occurrence. Why does the probability of occurrence receive much less attention?

Consider a potential tanker accident that would result in a larger oil-spill. The largest potential spill size is the total amount of oil carried by the tanker. The probability of occurrence of an accident is based on an analysis of data on tanker accidents, design of the tanker, and various safety devices or procedures that would reduce the likelihood of an accident. Clearly, the consequence (spill size) is easier to visualize than the probability of an accident. The latter is not so easy to comprehend, and there is general scepticism about predictions of the high reliability of technological systems. Murphy's Law is assumed to prevail here: if something can go wrong, it will go wrong, eventually.

Given this scepticism, many who may be affected by an oil-spill are reluctant to accept an accident probability as small as one in a million per year. They would be content if the probability were zero, but that is impossible.

Future EARP panels will still confront a sceptical public. Nevertheless, efforts ought to be made to educate the public that risk consists of two interconnected components, probability of occurrence and conse-quence. Perhaps future panels could develop guidelines for defining acceptable risks, so as to incorporate both probabilities and consequences.

For example, suppose that the impact of 160 cu m (1,000 barrels) of spilled oil on a species is considered "significant," or, in the words of the Beaufort EIS, a "major" impact. Would a reduction in spill volume to, say, 130 cu m change the impact from "major" to "moderate"? Would 120 cu m be "minor"? The EIS did not correlate volumes of spilled oil and impact, nor did the panel. Without such information, how can proponents decide on design changes to the arctic tanker, intended to reduce spill size?

The Beaufort EIS developed scenarios of hypothetical accidents re-sulting in oil-spills at various locations in the Arctic. Each accident affected local species, in impacts that could range from "negligible" to "major." If "major" impact is to be avoided, then arctic tankers should be designed so as to reduce a "major" impact to "moderate" or some other acceptable level. Then one species that is very vulnerable to spilled oil may govern tanker design. Whether this species is vitally important is not easy to determine, given the sparse knowledge of arctic ecosystems.

Because of the uncertainty about acceptable risk, future EARP panels may have to use hazards as surrogates. If one knows even roughly the relation of spilled-oil volumes to environmental effects, reducing the amount of spillable oil limits hazard (the spilled oil). In limiting hazard, panels should err on the side of caution.

CHAPTER TEN

Science Policy and Ocean Management
Douglas M. Johnston

This volume concentrates on recent ocean science and environmental assessment experience in the Canadian Arctic, rather than on arctic science or environmental assessment practice in general. The contributors have been asked to react to the prospect of increasing developmental intrusions in a sector of the natural (and human) environment that seems to require special protection. This chapter, however, reflects on science policy: how should Canadian science be asked to improve public (and possibly private) decision-making in design of transit management for northern waters?

DILEMMA AND PARADOX

As Lamson indicates in chapter 1, there are two clearly distinct approaches to arctic science. The "idealist" approach looks on the Arctic, like the Antarctic, as a "natural laboratory"; the "utilitarian" approach looks on the Arctic as a "strategic resource depot." Behind this distinction one can discern several layers of significance. At the first level, the "idealist" approach seems to assume the primacy of *values* (or a certain cluster of values), whereas the "utilitarian" seems to assume the primacy of interests (specifically those invested in the research). The altruistic, disinterested orientation of the "idealist" approach reflects the dignity, or even nobility, of the scientific tradition, characterized as a nearly selfless quest for knowledge for its own sake. It seems to call for an emphasis on "pure," as distinct from "applied," science. However, it reflects also a "classical" attitude to the acquisition of knowledge, which can be criticized as insensitive to more general and more basic human needs.

In this vein, the acquisitive, developmental thrust of the "utilitarian" approach offers science in a cause that can be related, more or less directly, to basic and non-elitist human needs. It calls for "applied" rather than "pure" science. Moreover, in an age when most arctic science is likely to be perceived as part of an expensive, large-scale investigation, the trend in Canada is to public funding based on fairly specific criteria of relevance, efficiency, productivity, and accountability. On first impression, then, the distinction between "idealist" and "utilitarian" reveals a dilemma between knowledge and public policy, but both contributing to the betterment of human welfare.

At a second level, the idealist approach holds out an image of the arctic region, or at least the Arctic Ocean, as an "international" area or as an area of special international significance. This internationalist argument can be supported by reference to the unique fauna and flora of the region, its unique role as regulator of the global climate, and its special suitability for a wide range of non-biological and non-climatological studies. The Arctic Ocean, like the antarctic region, can be viewed, in whole or part, as a "world heritage" site, based on criteria formulated under UNESCO's World Heritage Convention. In contrast, the utilitarian approach seems to rest more squarely on current political realities, such as the firmly established trend to extended national jurisdiction in the ocean and the lack of any exception for the Arctic Ocean. Yet, in light of technological developments geared to facilitating year-round navigation in the Arctic, the utilitarians' more nationalistic orientation to science policy must eventually be designed to serve international as well as national purposes.

In the mean time, as a matter of political realism, it must be assumed that most of the critical decisions both on arctic science policy and on ocean science policy will be made by national governments, mostly in consultation with the relevant industries and with inputs from northern communities through land use planning commissions (e.g. Lancaster Sound Regional Land Use Planning Commission). At that level, arctic ocean science policy is likely to reflect a heavy utilitarian bias. In Canada, however, research designed with a view to its future role as "manager" of the Northwest Passage must be influenced partly by its international responsibilities to foreign users as well as domestic residents of the Arctic, since this is inherent in any concept of "transit management."

At the time of writing, the relevant decision-making in Canada seems inhabited by a further paradox. Severe cut-backs in government spending require significant reduction in the scale of ocean science in Canada, and almost certainly a less ambitious program will result in some loss of quality. It must be doubted whether industry and the academic community can take up the slack, as apparently envisaged by the federal government. The profit-making priority in industry and the parity of teaching responsibilities in the universities constrain the private sector's capacity for quick and significant expansion of research effort, especially the arduous type conducted in northern waters. Yet, paradoxically, reduction in the scale of arctic ocean science coincides with recent government decisions to spend more than ever on arctic ocean technology, especially in the form of ice-breaking or ice-strengthened vessels for management of the Northwest Passage. To the extent that such technology seems to require more, rather than less, scientific research of certain kinds, arctic ocean science and technology must be designed in concert to match the government's right hand with its left.

SCIENCE IN THE SERVICE OF MANAGEMENT

For the specific, utilitarian purposes of transit management in northern waters, the design of science policy in Canada should pivot on the concept of "management." But precisely what kind or kinds of management should be involved? If science should be designed to serve management, how many different contexts of science and science policy should we look to for guidance? To reflect on these questions, it may be sufficient to refer to seven distinguishable kinds of management. Unlike so-called crisis management, each of these seven types represents a more or less systematically planned, scientifically based approach to the treatment of foreseeable problems.

Species Management

Originally, species management was designed and supported almost exclusively by biologists, and that tradition has been preserved virtually intact in the case of wildlife species of no commercial significance. Such species have proved amenable to a relatively uncomplicated and inexpensive form of management, even in highly mobile species such as migratory birds, which require international co-operation in the exchange of scientific data. Management of this "minimal" sort, focusing on preservation, involves occasional, wholly benevolent human interventions in nature.

Management is much more difficult, expensive, and controversial when the designated species is of commercial significance. Then management must be based on a more elusive concept of public policy, shaped by economic, industrial, communal, and therefore political, as well as biological, considerations. Relevant values and interests are likely to be found in conflict. The relevant disciplines, dominated by economics and biology, may be constantly locked in controversy over questions of stock conservation, rather than species preservation, and reconciling viewpoints may be impossible. Additional or more reliable data may not necessarily provide a more "rational" basis for management decisions of this sort, but at least the sharpness of the focus on species, rather than area, sometimes facilitates difficult decisions on the prohibition or strict restriction of commercial usage, as in the case of certain whales, seals, and polar bears.

With the introduction of "elite" tourism to arctic regions, the experience of species management may be of increasing relevance to the much larger task of designing a system of transit management. At present, however, more emphasis might be placed on population and behaviour studies, in light of the prospective threat of noise and pollution to vulnerable species of the arctic marine environment.

Habitat Management

Habitat management, dominated by biologists, is based on area rather than species, but the designated area is usually defined with reference to a seasonal activity such as breeding or spawning, which is confined to identifiable local sites. This kind of management tends to be efficient – and cost-effective – because it focuses on the most vulnerable stage of the life-cycle of a species. Benevolent human intervention can take a highly effective form, such as physical protection of the site from marauders, human or otherwise. Characteristically, management authority of this kind usually includes the power to exclude all non-scientific purposes and personnel in the area.

Since there are several alternative routes through the Northwest Passage, transit managers must be authorized to determine which routes are permitted during each season, in light of known seasonal movements of vulnerable species and changing seasonal significance of known sites as habitats for breeding, spawning, or other natural purposes. The experience of habitat management will be valuable for determining the most appropriate scientific basis for route selection on a month-to-month basis.

Ecosystem Management

Ecosystem management is newer than species or habitat management, and it is not yet known how broadly based it should be, for any given purpose. Since it is based on maintaining ecological balance, rather than preserving or conserving specific species or their habitat, this kind of management purports to apply to an area of interconnectedness that might be broadly, and somewhat arbitrarily, defined. To the extent that it depends on ecological rather than biological research, it might appear an expensive alternative to species or habitat management.

Its relevance may depend on Canadian ecologists deciding on the distinctness of the ecosystems in the arctic marine environment, or at least sectors likely to be affected by increased vessel traffic in the next two or three decades.

Resource Area Management

Resource management often represents a sectoral and, in some degree, industrial perspective on ocean use and management. It is much more limited than environmental, or even ecosystem management. But resource management based on an area, not a species, must consider other sectoral, or cross-sectoral, issues arising in management. To the extent that resource management is associated with industrial considerations, scientists involved must find a way of accommodating their values, methods, and data with those of economists and others trained in institutional concepts and arrangements. Scientific concern with biological conservation must be reconciled with avoidance of waste and inefficiency. In these respects, resource management experience seems relevant to transit management in the Arctic.

Recent experience in the management of ocean fisheries looks particularly useful, since most fishery resources in the world's oceans now come under the authority of a managing coastal state, just as the Northwest Passage comes under Canada's authority. There is a growing world-wide trend to an increasingly eclectic conception of efficient and equitable management for ocean fisheries, whereby managers regard their work as more political than scientific. Not least in Canada, fishery management now seeks some kind of sub-optimal "rationality" that might satisfy the resource industry, coastal communities, and the world of science. Recent studies of Canadian fishery management clearly reflect this view.

However, industrial fishery development and management are not likely to have high priority in the Northwest Passage, though they may in the adjacent Davis Strait. Navigation, moreover, is not a resource. In transit management in the Arctic, fishery specialists might do more harm than good, tempted for the sake of political peace to make the wrong kinds of compromises among industry, government, and the local (Inuit) communities. Experience with fishery management might even have bred scepticism, bordering on cynicism, regarding the reliability of scientific data as the basis of management decisions.

Use Management

It might seem safer to draw instead on the experience of those involved in management of a specified use, especially of an ocean area, as in vessel traffic management. However, this field draws chiefly on the research findings of social and environmental impact assessment, which have recently been severely criticized for a variety of reasons. This kind of science is often too hasty, and sometimes prejudicial. Almost invariably, at least in Canada, its influence is limited to the preparatory, pre-approval period. More often than not, it is ill-suited as a long-term basis for use management. Too frequently, advice in this form can be easily dismissed as coloured by the scientist's social judgment or political opinion, even though scientific findings that lack objectivity may still be valid.

Given the deficiencies of impact assessment for use management, management decisions of this kind may be taken with inadequate, or even minimal, constraints from the scientific community. This approach may be the most serious threat of all to the design of an effective system of arctic transit management. A use-specific approach to transit management sounds economically sensible and cost-effective, but may not be appropriate, given the uniqueness and complexity of such a difficult environment. Perhaps Canada cannot afford to develop a broad scientific basis of completely reliable data. It would be foolish to entertain ambitions that exceed one's grasp. Yet it may not be wise to sacrifice breadth in the hope of promoting reliability.

Technology Management

The argument for specificity of approach seems more cogent if the focus is transferred from management of use or activity to management of technologies. Emphasis on technology is compatible with emphasis on the techniques of management: innovations may provide the best way to improve the management system. Technology management rests on science in the form of assessment of technology, risk, and impact. It is broader based than use management, which relies on impact assessment. Because high technology has a built-in dynamic, prescribing constant change and improvement, technology management will stress continuous research and development.

Applied to transit management in the Arctic, the technological approach encompasses navigational and communications components of an ultra-modern transit system for safe passage in a remote and potentially dangerous environment. Use of the most modern technology, backed up by appropriate science, would certainly reassure nervous users and thereby raise the revenues available for management in general and for the appropriate science in particular.

Decision Management

Ultimately, however, the chief objects of management are people. A democratic society as open as Canada's requires an open, participatory process of management. Much has been said about effective partnership between federal and provincial (or territorial) governments and between government and industry, and – in resource-related contexts – about involving local communities in management decision-making. An expanded social role for the university community has also been advocated. In a highly complex situation, the management "system" or "process" is, or should be, a network of communications. Management becomes not only the final, formal decisions of the professional managers, or their institutional apparatus, but the whole network.

The input from science is the total flow of information, intelligence, and imagination presented to professional managers through the network. There should be professional and non-professional "managers" and "scientists." Increasingly, a "scientific" adviser is someone with easy access to a data bank, with the education and intelligence to interpret relevant data. A non-professional "scientist" may lack formal university education in science and yet contribute to management as an acutely perceptive observer.

This broad conception of science and management may offer an appropriate and cost-effective way of broadening scientific input into a highly complex sector of public policy and administration.

CONCLUSION

As VanderZwaag suggests in chapter 11, any major issue over use of Canadian arctic waters is likely to reflect deep-seated tension between immanent and transcendent attitudes to nature. For something as complex as a complete transit management system for such an environment, scientific input must be designed in a broad, imaginative, and relatively ambitious manner. A "rational" basis for management can hardly be narrow, unimaginative, or timid. In these respects, reason seems contingent on "transcendent" policy-making attitudes and initiatives but should be tempered by the manager's sensitivity to humankind's "immanence" in nature.

Both kinds of attitudes to nature possess their own kind of legitimacy, even integrity. Surely it is possible to reconcile ethical and political considerations through an appropriately designed system that draws on a wide range of science-based experience in ocean development and management.

What lessons for public policy and management should we learn from Canada's recent experience in arctic ocean science? This is the question underlying all the chapters in this work, and one that the Canadian public must ultimately address.

CHAPTER ELEVEN

On the Road to Kingdom Come
David L. VanderZwaag

Harry Chapin, the late folksinger, perhaps captured the essence of human existence in "On the Road to Kingdom Come." The song describes the human predicament of continually seeking and expecting perfection or absolute happiness but finding disillusionment and dissatisfaction instead. The leader, who hopes for great power to change the world, discovers that his freedom is shackled by traditional bureaucracies and an uninventive public. The general, who looks forward to a peaceful and enjoyable retirement, finds instead loss of physical activity and the disappearance of respect and power. The old man, who purchases cologne and candy in expectations of creating a wild romance, experiences only a momentary flash of excitement. They/we keep establishing new goals, new expectations, new ideas, to escape the real predicament: "We are all travellers on the road to Kingdom Come."

Like it or not, scientists, environmental assessors, and decision-makers also operate in the real world and also run head on into human limitations and human dissatisfactions, the parameters of which have been touched on by the previous authors. Scientists would like to be "objective," you might even say "absolutely perfect," by being able to predict unreservedly (or with few reservations): "X oil spill will cause y and z effects on the environment." Risk assessors would like to be quantitatively "objective" by saying: "The risk of an oil spill from an arctic class-7 tanker is one in one million and in case of such an oil spill the ecological effects will be a loss of x number of seabirds and a 10 per cent reduction in recruitment of certain marine mammals." Decision-makers would also like to be "objective" by confidently declaring: "This shipping proposal is [or is not] environmentally and socially acceptable."

But as the environmental policy literature makes abundantly clear, "uncertainty" and "subjectivity" enter all our decision-making processes. Scientists have not conquered all the complexities of ecosystem relationships and are often forced to fashion artificial boundaries and to rely on "best guesses" (Edmunds 1981). Risk assessors have recognized the great gulf between their quantified assessments of risk and public perceptions of risk (Douglas and Wildavsky 1982; Slovic, Fischhoff, and Lichtenstein 1980). Decision-makers, cut adrift without legislative definitions of "acceptable" risks, realize that the answer depends ultimately on value preferences and personal interests. Thus scientists, assessors, and decision-makers are all travellers on the road to Kingdom Come, hoping and looking for new solutions, new approaches, or worse, resolved to bureaucratic disillusionment.

Numerous "signposts" to Kingdom Come for the arctic offshore, that is, ideas and arrangements for fairer and more stable decision-making processes, do exist or are emerging. The ethical reality of conflicting world-views and value premises and varying risk perceptions among experts, the public, and local communities point to participatory, consensus-seeking decision-making processes. Scientists have suggested ways to improve scientific inputs into the Environmental Assessment Review Process (EARP). Academics and policy analysts have argued for procedural and substantive strengthenings of the assessment process. New mechanisms for potentially lessening conflicts over offshore resource use have emerged, including northern land use planning and wildlife management boards and an environmental impact screening committee created by the Committee for Original Peoples' Entitlement (COPE) land claims agreement in the western Arctic. Ongoing negotiations between the Tungavik Federation of Nunavut and the federal government over land claims in the central and eastern Arctic may also result in new forms of co-operative decision-making and even of revenue-sharing.

This chapter briefly examines the signposts in a six-part discussion: ethical considerations, realities of risk assessment, scientific improvements in environmental assessment, procedural and substantive strengthenings for environmental assessment, new directions in northern decision-making, and conclusion.

ETHICAL CONSIDERATIONS

What is a good environmental assessment? What is a good decision-making system? Putting aside the objective importance of securing sound scientific information, the answer will depend ultimately on two subjective factors. The first factor is the respondent's personal interest. A person owning recreational property or a person solely reliant on renewable resource harvesting, faced with a major industrial proposal for his or her backyard, would likely respond, "A system, an assessment that stops the project as soon as possible and at the least cost to me."

The second factor is the respondent's philosophical world-view or value priority system. As described by numerous authors, there tend to be two polarities in ethical views toward nature – transcendent and immanent (Tribe 1974), also called consumer and exerter (Gibson 1975), interventionist and gaianist (O'Riordan 1986), or materialist and post-materialist (Watts and Wandesforde-Smith 1981). People drawn toward the transcendent pole tend to view humans as above or dominant over nature. Their overall goal is to maximize the satisfaction of human wants, with nature as the object and provider of resources. Characteristics such as risk-taking, competitiveness, centralization, efficiency, and resource development are preferred values. Persons drawn toward the immanent pole view humans as part of nature, and their major goal is maximizing the protection of nature, since natural things have their own subjective, intrinsic value. Such persons tend to value or favour risk-adverseness, co-operation, decentralization, equity, and traditional ways.

Again, the vision of the good decision-making system may be coloured by one's values, with the transcendent viewpoint seeking "the system that maximizes resource exploitation" against the immanent viewpoint, "the system maximizing the protection of nature."

Of course, conflicts of values and interests are far from simple: people do not come in neat psychological or philosophical packages, and there may exist a spectrum of ideologies. Value conflicts do not just occur between persons but also within persons (Sagoff 1982). For example, I may oppose offshore drilling on moral grounds, in that the environmental risks are too great, but at the same time support drilling because of consumer interests, such as the need for a secure gas supply for my two cars. Even business leaders, who tend to be transcendent in world-view and who favour risk-taking, competitiveness, and market controls, do not always rank economic growth as a priority over environmental protection. A study of value priorities in the United States, England, and West Germany showed that a high percentage of business leaders (e.g. 30 per cent in the United States, 37 per cent in England) were neutral, and many favoured environmental protection (e.g. 29 per cent in the United States, 48 per cent in England) (Milbrath 1981). Tribe (1974), in his article "Ways Not to Think about Plastic Trees: New Foundation for Environmental Law," saw the dangers in becoming cemented to either pole. The fully transcendent world-view may lead toward a world of plastic trees, while the totally immanent view may lead to intolerance, fanaticism, and economic/social stagnation.

What does this ethical reality have to do with environmental assessment and northern decision-making? The ethical reality may assist us in perceiving and understanding five dimensions of decision-making: the intertwining of form (decision-making structure) and substance (interest and value positions); the bias toward transcendence inherent in the existing system; the tradition of fragmentation and "ad hocery" in northern decision-making; the need for a new philosophy underlying our decision-system; and the need for new directions in the decision-making process.

Intertwining Form and Substance

Different views as to substance, that is, what is a good decision based on value or interest assumptions, will likely lead to different views as to ideal decision-making frameworks. Both Inuit, who generally hold a more immanent world-view, and industrial proponents, who generally favour a more transcendent world-view, would probably prefer an authoritarian system of decision-making, favouring their particular philosophical position. An authority-model structure may include rigid statutory standards (either favouring or disfavouring environmental protection), broad discretion to exempt projects and programs from general laws, and designation of a particular agency with broad override powers (Andrews 1981). An example of this tendency to favour a particular structure is provided in the chapters by Dryden (5) and Jull (6). Dryden speaks of the desire for a single-window approval system and a joint government-industry working group to focus on impact prediction, mitigation, and monitoring. Jull speaks of the Inuit desire for their own public institutions and jurisdictional clout to manage their own society.

Examples of the authority-model in operation include Canada's Northern Pipeline Agency, created in 1978. It was intended to "fast track" the building of the Alaska Highway Natural Gas Pipeline. The agency was delegated powers from the National Energy Board and the departments of Environment and of Fisheries and Oceans, including the right unilaterally to set environmental conditions (VanderZwaag and Lamson 1986). Industry might point to various discretionary land withdrawals pursuant to the Territorial Lands Act. An example from US experience is the Endangered Species Act, which at one time provided "absolute" protection for endangered species. When a project such as the Tennessee Valley Authority (TVA) dam confronted an endangered species, even the lowly snail darter, the project could be prohibited. Subsequent amendments "watered down" the legislation by providing for project exemptions under narrow circumstances, including situations where there are no reasonable and prudent alternatives and the action proposed is in the public interest.

If clear value preferences are not given or achievable, then second-best structural options may be preferred. An environmentalist would probably favour a complex decision-making system involving numerous approval points in order to get as many "kicks at the project can" as possible. Adversary litigation might offer an additional tool for delaying and for raising the political profile of the issue. An industrialist might prefer broad official discretion and procedural simplicity, so that quick decisions might be attained through political lobbying.

When decision-making structures become the whipping post of industry, native groups, and public interest groups, the ethical reality bids one to examine the underlying value positions and interests snapping the whip in order to grasp the substantive issues.

Bias in Decision-Making

The existing legal framework for managing the arctic offshore contains many aspects that tend to favour the transcendent paradigm. The pro-development slant begins in the common law, where an Inuit plaintiff seeking to challenge a pollution activity, such as disposal of offshore drilling fluids or tailings from a lead-zinc mine, would bear the burden of proving damages and might also have to show legal standing (a special interest such as a property right or personal injury).

The legislative regime is nearly devoid of procedural rights and substantive decision-making powers for northerners. For example, the Fisheries Act and the Arctic Waters Pollution Prevention Act (AWPPA), which provide power for extensive departmental reviews of projects proposing to deposit deleterious substances or wastes, do not guarantee public input into the decision-making process through public notice of reviews, public consultations, or public hearings. Neither act requires Inuit involvement in decisions concerning vessel routing and establishment of exclusion zones. The Arctic Shipping Control Authority and the Environmental Advisory Committee on Arctic Marine Transportation (EACAMT) evolved out of the EARP report for the Arctic Pilot Project (APP) and operate without legislative guidance concerning funding and Inuit participatory rights (although Inuit representatives have been included as members of EACAMT). The land use planning process, unless results of land claims dictate otherwise, will continue to operate without a legislative umbrella and may remain largely an advisory forum. Both the Fisheries Act and the AWPPA allow pollution if granted governmental blessing through regulations or permits and therefore are utilitarian, not utopian. Pollution is rationalized and, after the necessary permit is issued, legalized (Emond 1984). EARP continues to operate without the clear authority of legislation, and panel recommendations are binding only by ministerial discretion.

The list could go on, yet not all aspects of the legal framework work or appear to work against environmental protection. Provisions for broad official discretion, without limiting time frames or detailed procedures, and duplications in review processes may also be tools for delaying beneficial projects and may prove unfavourable to industrial proponents.

Fragmentation and "Ad Hocery"

No one is likely to disagree that the decision-making processes surrounding northern resource development have been fragmented and ad hoc. At least thirteen different federal departments and agencies have jurisdictional threads of control over the arctic offshore, including the Canada Oil and Gas Lands Administration, over hydrocarbon exploration/exploitation; the Department of Indian Affairs and Northern Development (DIAND), over resource development and environmental protection; Transport Canada/Coast Guard, over marine transportation; the Department of Fisheries and Oceans, over renewable resource harvesting and marine environmental protection; and the Department of Environment, over ocean dumping and marine environmental quality.

As noted by Dryden, in chapter 5, tremendous inconsistencies have occurred in governmental reviews, with many northern transportation schemes not subject to a full or even a superficial environmental assessment. Major mining projects such as the Nanisivik and the Polaris mines avoided formal EARP review. Offshore drilling at the exploration phase in the Beaufort Sea and High Arctic islands escaped the extensive environmental assessment engulfing the APP.

Ethical considerations may shed some light on the reason for such a confusing, tangled system of offshore management. In a world of competing values, where no consensus exists on the type and direction of society, government leaders tend to avoid clear statements of policy and tend to construct numerous administrative buffer points for defusing political voices and accommodating conflicts of interests and values. Layer upon layer of sectoral divisions of bureaucratic turf has been built up through the years with little concern for integration or co-ordination. Given this ethical and political reality, rational ocean management may be beyond our grasp.

A New Philosophy

If one accepts the negative implications of unlimited growth and no growth, that is, the transcendent and immanent ethical polarities, then the need for another guiding philosophy – something between the two poles – arises. That something has been given various labels, including balanced development, sustainable development (Environment Canada 1984), and eco-development (Loubser 1984). There are likely no agreed definitions, but a number of characteristics seem common: favouring balanced economic growth on a human scale and pace, passing on a viable resource base and environment to future generations, promoting development of renewable resources, and avoiding rules and relations instituted at the highest level in favour of participatory planning.

Some might argue that Canada, through creative use of EARP, has stumbled toward the eco-development paradigm. For example, the Beaufort EARP report favoured a small-scale pipeline for transporting offshore hydrocarbons, recommended against shipping until further testing was carried out and government preparedness improved, emphasized involving local residents in land use planning and port management, and recommended further transfer of political control to the territories (FEARO 1984).

However, as the chapter (6) by Jull emphasized, the Inuit still view EARP as "a tool in the service of industrialization," and, as Mr. Justice Thomas Berger has noted, the fleshing out of the eco-development philosophy will not occur overnight but will involve a lengthy process: "It is rather the rational application of industry and technology that we must pursue. But we cannot expect that within a week, or a month, or a year, a new philosophy can be worked out in all its details ... We must realize that if we are to postulate, let alone erect, an alternative to a system established 400 years ago, and which has ramified throughout the world, we must be prepared to begin on a small scale. Small can be beautiful, and that applies to theorizing as much as to anything else" (Berger 1984).

New Approaches

The ethical reality of conflicting interests and world-views, while not providing an "instant fix" or concrete image of the perfect decision-making process, does offer a few conceptual lenses for perceiving the direction the process should take.

First, the need to take all value positions seriously calls for minimal reliance on the authoritative model of decision-making. A process involving strict statutory standards, project exemptions, or a single mission agency with override powers tends to foreclose value analysis and discussion in favour of making one position paramount (Andrews 1981). Since so many environmental questions involve scientific uncertainty, varying interests, and value conflicts, negotiation involving creative bargaining seems more attuned to resolving conflicts (Dorcey 1986). However, the authoritative approach may still play an important role – for example, in setting strict environmental quality standards after scientific indications of environmental harm or following a fair bargaining process. The approach may also be necessary for setting aside areas of special ecological significance, in providing special protection to rare or endangered species, and in resolving non-negotiable value conflicts.

Second, the ethical reality urges policy-makers to recognize the limitations of the scientific model of decision-making, which may involve the assumption by officials that the public interest can be discerned "objectively" through such tools as cost-benefit analysis, public surveys, quantitative computer modelling, systems analysis, environmental impact statements, and coastal-zone management plans (Andrews 1981). Such tools may be useful in identifying and mitigating environmental and social affects, but more empirical data and plans do not necessarily mean better decisions, for public legitimacy may still be lacking, public perceptions may be beyond measurement, and not all values can be quantified (Andrews 1981).

Third, ethical considerations encourage participatory democracy in decision-making (Gibson 1975), which calls for full public involvement in all phases and aspects of decision-making, "for we are all experts in value judgments" (Elder 1975). Direct debate among industrial proponents, government officials, and public interest groups may occur through negotiation, collaborative planning, and mediation, with the ultimate goal of compromise and mutual accommodation (Andrews 1981). Participation has been seen as a keystone of environmental protection by many writers (Elder 1975; Gibson 1975; Morley 1975; Emond 1984; and Dorcey 1986), perhaps because it symbolizes so many things – a change in power relations from domination to co-operation, a method for making administrators aware of public views, a form of conflict resolution through increased understanding and tolerance by means of sharing views, and a system of social therapy through reduction of public alienation (Wengert 1985). A consensus-seeking style also seems more compatible with Inuit collective and co-operative social structures (Whittington 1985).

REALITIES OF RISK ASSESSMENT

Although quantification of risk may assist decision-making by providing a concrete image of the chances of a pollution incident occurring and the potential extent of damages, risk assessment techniques do not "objectively" measure acceptability, for subjectivity may enter in through an expert's speculative filling of knowledge gaps and through varying public perceptions. As Lemberg noted in chapter 9, estimates of probability for an arctic shipping accident suffer because of lack of experience with commercial transport of oil and liquefied natural gas (LNG) under arctic conditions, and estimates of damage are open to guesswork due to local variability in wind, wave, ice, and current conditions, uncertainty about available counter-measures, and lack of knowledge about cumulative environmental effects of hydrocarbon spills. The public may perceive the risks from offshore development as much greater than the quantified reality of the experts. Potentially catastrophic and involuntary risks tend to be psychologically magnified (Smalley 1980), and public perceptions may be coloured by beliefs much broader than the idea of personal danger: beliefs that a project has no overall economic benefit and is not needed, will make society more consumer-oriented, or will further huge industrial enterprises (Thomas and Otway 1980).

As with ethical considerations, the subjective realities of risk assessment indicate directions for the decision-making process. Lemberg (chapter 9) points out an educative direction: the public needs to be educated about the benefits and limits of quantifying risk. He also pointed to a definitional direction. Environmental assessment panels, in fairness to a proponent of development, should define acceptable risk – for example, by establishing hazard limits such as a certain volume of oil spilled or by focusing on the protection of certain valued species. The psychological realness of public perceptions of risk points to a participatory direction. The public must be provided a fair opportunity to express perceptions, which should be taken seriously and heeded by decision-makers. The psychological magnification of involuntarily imposed risks points to a beneficial direction. People are more likely to accept a risk if there is some voluntarily accepted benefit, such as revenue-sharing (something the Inuit and northern territories have been anxious to obtain) compensation, or a more creative form of benefit, such as a corporate financial contribution to an environmental cause.

The importance of this beneficial direction was brought home to me by the APP. In case the Nova Scotian option for a southern terminal was chosen over the Quebec location, I would have been a resident in the potential "catastrophe" zone. Class-7 LNG tankers would probably have docked 8–10 km from a wilderness island off the Strait of Canso where my wife and I sojourned for a few years and lived in harmony with abundant wildlife, including deer, lynx, osprey, and eagles. We in no way would have found the APP acceptable, even given a low probability of accident. We and our fellow living creatures were the potential personal losers, not the corporate executive in Calgary or the bureaucrat in Ottawa. However, if the proponents had considered some innovative approaches, such as financial assistance in protecting some of the surrounding islands, perhaps through some form of land trust, our attitude would have been more positive. We then would have perceived some benefit to be weighed against the involuntary risk.

The Beaufort Sea EARP Panel also perceived the close connection between risks and benefits: "Collective risks to the renewable resource base of northern residents must be offset by significantly increased local, northern benefits. The establishment of a Northern Heritage Fund has been suggested by the Senate Committee on Northern Pipelines. The Panel supports the creation of such a fund as an interim measure. This would allow negotiations on revenue sharing to proceed concurrently with any part of the development proposal."

SCIENTIFIC IMPROVEMENTS

The scientific and decision-making perspectives expressed in this book and the growing environmental assessment literature point to at least nine directions for improving scientific input into the environmental assessment process.

The chapters by Lake (2), Smiley (3), and Brown (4) emphasize the informational direction. Numerous informational gaps in scientific knowledge need to be filled by additional basic research. Brown notes that while we have reasonably adequate information on immediate environmental risks to seabirds, we know little about long-term risks, such as the effects of oil ingestion by seabirds, and even the short-term effects of vessel noise and sea-ice disturbances on seabirds. Lake highlights greater understanding of atmosphere-ice-ocean interactions in the marginal ice zone, in order to increase weather- and ice-forecasting capabilities. He also speaks about improving ocean-measurement technologies, particularly in the Arctic, where frigid temperatures make equipment failures common. Smiley discusses our limited understanding of the effects of ship noise and vessel tracks on marine mammal behavior. The Beaufort Sea EARP Panel recommended additional research into the effects of offshore development on seals, bowhead and beluga whales, narwhals, seabirds, and fish and ways to minimize effects on wildlife in the Cape Bathurst and eastern Lancaster Sound polynyas.

Some authors and the assessment literature emphasize a temporal direction, expanding the time frame of scientific research. Lake criticizes the seasonal bias of most data, since the great majority of arctic ocean-ographic research takes place during the "fair-weather" months of spring and summer. Brown regrets that so many research programs, such as the Eastern Arctic Marine Environmental Studies (EAMES), which are mission-specific and operate only for a few years, under perhaps highly aberrant environmental conditions, produce impressionistic "snap-shots," not holistic, ecological "movies." Marshall (chapter 8) expresses concern about major government funding for industrial devel-opment (for example, the billions of dollars spent in the petroleum incentives program) without a corresponding commitment to long-term research. Others have urged environmental studies much earlier in the assessment process, with economic and engineering feasibility studies (Rosenberg et al 1981).

Some bright lights, advocating long-term research programs, dot the northern policy-making horizon. The Beaufort Sea Panel recommended a fifteen-year program of accelerated arctic research and a federal arctic research policy, including a mechanism for funding long-term research. The Federal Environmental Assessment Review Office (FEARO) has es-tablished the Canadian Environmental Assessment Research Council (CEARC) to promote long-term research in the physical, biological, and social sciences in order to improve the scientific foundation of envi-ronmental impact assessment (EIA) (Hanson 1985). The council has iden-tified four major research themes: integrating impact assessment with strategic or regional planning, improving application of ecological and social sciences to impact analysis, improving procedures for clarifying and incorporating social values in impact evaluation, and examining alternative policy and institutional frameworks (CEARC 1986). The Inter-agency Arctic Research Policy Committee (IARPC) in the United States has developed a five-year plan for arctic marine ecosystem research which emphasizes co-operation with other interested nations (IARPC 1987). Scientists and scientific administrators from eight arctic nations (Canada, Denmark/Greenland, Finland, Iceland, Norway, Sweden, the Soviet Union, and the United States) met in Stockholm in March 1988 and agreed that an International Arctic Science Committee should be established to promote international co-operation and co-ordination of scientific research in the Arctic (IARPC 1988). Canada is considering establishment of a Canadian Polar Research Commission and a Polar House to promote and co-ordinate polar research.

However, the Canadian government's commitment to long-range scientific research in the Arctic remains open to question because of financial restraint. Funding for the Northern Oil and Gas Action Pro-gram (NOGAP), a major research program to prepare for offshore hydrocarbon development, was cut by over 40 per cent (Faulkner 1985; Tener 1985). The Department of Fisheries and Oceans was ordered to trim $25 million from its 1986 budget, and some scientists feared termination of a number of ocean science programs in Canada (*Globe and Mail*, 26 July 1985).

The dynamic nature of ocean phenomena and the evolving legal regime point to a spatial direction. Scientific research must move beyond artificial geographical limitations imposed by national/political boundaries or by narrow terms of reference for a project assessment, in order to match ecosystem realities. Transboundary ocean currents can often transport pollutants from one country or jurisdiction to another, and living resources (for example, bowhead and beluga whales) often undergo extensive migrations. An example of spatial limitation was demonstrated by the Beaufort Sea EARP, where research and review focused on issues north of 60 degrees. Recognizing potential tanker traffic south of 60 degrees, the panel recommended further environmental and socio-economic review concerning the effects of tanker traffic in the Labrador Sea. Both national and international legal frameworks call for scientific research to cross national boundaries. Section 4 of the EARP Guidelines Order requires a proponent of development to consider potential environmental and social effects external to Canadian territory. The Law of the Sea Convention (in articles 200, 242, and 243) calls on states to promote co-operation and integration of marine scientific research efforts.

Numerous scientists emphasize focusing of EIAs on key issues, so that information-gathering and prediction are manageable and financially realistic. Although public involvement, in identifying particular environmental concerns to be addressed by an environmental impact statement (EIS), is not required by the Guidelines, "social scoping" is seen as essential (Beanlands and Duinker 1983) and is becoming an accepted practice. FEARO held an initial scoping seminar in Calgary on 13 November 1980, to identify and narrow the environmental and socio-economic issues to be studied by the proponents of Beaufort Sea offshore development. Those present included representatives from industry, territorial and federal governments, native groups, northern communities, and special interest groups. For a proposed second nuclear reactor at Point Lepreau in New Brunswick, the EARP panel invited members of the public to scoping workshops, in order to formulate the guidelines for the EIS (VanderZwaag et al 1986).

Environmental assessment should be a continuing process of monitoring and learning from implemented projects (Roots 1986), not a once-and-for-all abstract study. "Learning by doing" was advocated by both the APP and Beaufort panels, through support for field trials of small-scale arctic shipping, and now has a legal base. Section 33 of the EARP Guidelines Order requires environmental monitoring programs for approved projects.

The other directions for the use of science in EIA are ecological: heavier reliance on field testing rather than on limited laboratory testing (Cairns 1986) or theoretical speculation; technical: clearly distinguishing technical, scientific questions from social and environmental policy questions; regional, or cumulative: looking beyond the effects of site-specific projects to potential cumulative effects of a number of projects over an entire region (Erckmann 1986); and planning: placing EIA within the context of a planning process (O'Riordan and Sewell 1981; D.A. Munro 1986), such as land use planning or coastal-zone management planning, so that social goals and long-term development strategies may be clarified.

CEARC has designated cumulative effects assessment (CEA) as a priority and has recommended specific steps to promote this type of assessment. The council is preparing to publish a reference manual on CEA methods, to encourage cumulative assessment of current and future projects and to promote a case study of cumulative effects in a prairie wetland/ agriculture situation involving three prairie provinces. The council also wishes to encourage research into all areas of CEA-theoretical development, scientific and methodological analysis, and institutional analysis (CEARC 1988).

STRENGTHENING ENVIRONMENTAL ASSESSMENT

Criticizing the federal EARP is somewhat akin to bad-mouthing a Good Samaritan, for EARP has played numerous positive roles in northern decision-making, most important, perhaps, filling vacuums in government planning and institutional preparedness. The EARP panel addressing the Norlands proposal to drill a single exploratory well in Lancaster Sound refused to approve drilling unless a regional planning process agreed on the "best use(s)" for the region, and it urged major expansion of government science programs to answer such problems as the fate and effects of oil from an offshore blow-out (FEARO 1979a). The APP Panel called for new institutions to manage shipping in the Northwest Passage, a control authority to regulate ship movements, and an environmental advisory committee to recommend shipping-related studies (FEARO 1980). The Beaufort Sea Panel's report reads like a "how-to" book on northern development, by listing eighty-three things that governments or proponents should do, from researching oil-spill clean-up equipment and building a class-8 ice-breaker to studying the feasibility of establishing higher education facilities at Inuvik and in the eastern Arctic (FEARO 1984).

Nevertheless, the federal assessment process has displayed at least eight major procedural and substantive weakenesses, which invite correction.

Toward Program Assessment

The present federal environmental assessment process does not clearly apply to government programs and related legislative proposals. For example, if the federal government were to announce a new Northwest Passage Transportation Program, providing a billion-dollar commitment to develop arctic shipping and management capabilities and a new control authority with broad approval powers, the program would probably not, based on past experience, fall under EARP scrutiny. Examples of government programs avoiding environmental assessment include the 1980 National Energy Program, which promoted offshore hydrocarbon exploration through petroleum incentive grants, and the recently announced federal program for Atlantic regional development, which provides financial incentives to businesses choosing to locate in Atlantic Canada. The EARP Guidelines Order leaves some ambiguity as to coverage, because the definition of proposals subject to the process does not speak of programs, policies, and legislative proposals: " 'proposal' includes any initiative, undertaking or activity for which the Government of Canada has a decision-making responsibility."

While one could argue that Parliament and its legislative committee system are a more proper forum for debate on environmental policy implications of government programs and legislation, much can be said in favour of an independent and perhaps supplementary environmental assessment. The legislative arena may not be fully sensitive to environmental values. Legislators, very often successful business persons, may tend to favour "big business" and economic growth. To stay elected, they may look only for short-term, easy solutions and may blindly follow party politics. Once the purse strings of government have been opened in a particular programmatic direction, assessment of individual projects may become superficial, merely giving mitigative credence to preordained development.

A number of Canadian task forces have recommended extending EIA to include government programs and policies. The federal Task Force on Environmental Impact Policy and Procedure, formed in 1972 to recommend an initial framework for the federal EARP, would have included as federal action subject to review "any policy, legislative proposal, program, project, or operational practice." The EIA Task Force of the Canadian Council of Resource and Environment Ministers (CCREM) stated: "Considerable evidence exists that suggests that E.I.A. processes which focus exclusively at the project level of analysis are incapable of dealing with fundamental environmental issues surrounding the need for the project in the public interest. To overcome this basic weakness in present E.I.A. processes, it is *recommended* that each government review the application of E.I.A. processes with a view to modifying them so that E.I.A.'s may be undertaken at both the program and project levels of analysis" (CCREM 1978).

Toward Independent Assessment

As provided in section 3 of the EARP Guidelines Order, the federal EARP leaves decisions concerning environmental significance and submission of projects to formal public review largely with the initiating department, through internal self-assessment: "The process shall be a self assessment process under which the initiating department shall, as early in the planning process as possible and before irreversible decisions are taken, ensure that the environmental implications of all proposals for which it has decision-making authority are fully considered and where the implications are significant, refer the proposal to the Minister for public review by a panel."

The conflict of interest inherent in self-assessment (Emond 1978) has probably contributed to numerous projects not being subjected to formal EARP review (Emond 1983). For example, exploratory drilling in the Beaufort Sea, begun in 1972 and involving 130 wells by 1981, avoided EARP (Rees 1984), as did exploratory drilling programs in the High Arctic islands and off the Canadian east coast, including the Grand Banks off Newfoundland and the Scotian Shelf off Nova Scotia.

The EARP Guidelines Order of June 1984 did grant FEARO supervisory powers over the initial screening and evaluation stages. Initiating departments are required to develop in co-operation with FEARO a list of proposal-types automatically excluded from the assessment process (section 11[a]), a list of proposal-types automatically referred to public review (section 11[b]), and written procedures for determining significance (section 16). The initiating departments must also report to FEARO on progress in implementing the review process (section 16), and FEARO is to make information on implementation public (section 18[d]).

Such arrangements may avoid past problems of initiating departments circumventing full EARP scrutiny, but independent screening and initial evaluation by FEARO should still be considered. The "FEARO farmer," not the self-interested departmental "fox," would be seen as guarding the environmental "chicken coop." More centralized screening and evaluation might make assessments more consistent.

At a National Consultation Workshop on Federal Environmental Assessment Reform, held in Ottawa, 10–12 March 1988 and drawing together 133 people from various backgrounds and all regions of Canada, participants proposed additional checks and balances at the self-assessment stage. They included an appeal process (to the minister of the environment, to an independent body, or to the minister responsible for a project) for contested self-assessment decisions, independent auditing to ensure self-assessment compliance, and preparation of an annual EARP implementation report (FEARO 1988a).

Increasing Public Involvement

The present federal EARP has rather minimal requirements for public involvement. While public hearings are required when a proposal threatens significant environmental effects or raises major public concern, public participation is narrow, involving short oral or written presentations to a formal panel, which then undertakes "black box" decision-making. Critical analyses, trade-offs, and final recommendations occur far removed from those directly affected (Wondolleck 1985).

Although the public may be involved in "scoping" exercises, as in the Beaufort and LePreau II projects, its involvement is not mandated in guidelines preparation. Section 30(2) of the EARP Guidelines Order provides merely that a panel may consult the public in preparing guidelines for an EIS. No public involvement is provided for at the screening stage. Even though the Beaufort Sea Panel provided intervenor funding of just over $1 million to facilitate public involvement, intervenor funding for EARP is not legally required and depends on ministerial discretion. No public involvement is assured after panel recommendations and project approval. Section 33(1)(d) of the Guidelines Order, requiring post-assessment monitoring, is silent about the role of the public.

Many authors have pointed to the above weaknesses and have recommended greater public involvement. Instead of a formal public hearing, where people present their concerns and complaints to an unpredictable panel, the assessment process might include a negotiation or mediation phase, where interested parties would seek consensus on likely effects, mitigation, and compensation and formulate reasons for unresolved differences (Dorcey 1986). The panel would then be presented with a statement of agreements. Negotiation might also be extended to other phases of EIA, including screening and developing research strategies (Dorcey 1986). To allay public misapprehensions about an approved project, local residents might be guaranteed involvement in project monitoring, perhaps through a community impact committee (Bankes and Thompson 1980; Lang 1981). The committee might watch for social and environmental effects, advise on necessary research, and recommend further mitigations.

Funding of intervenors should be legally required, and the rules for granting financial assistance to the public formalized, perhaps along the lines of the five criteria suggested by Mr. Justice Berger in the Mackenzie Valley Pipeline Inquiry (Lucas and McCallum 1975; Emond 1978): clearly ascertainable interest, presentation of a necessary and substantial contribution, an established record of concern and commitment to the interest sought to be represented, lack of financial resources to provide adequate representation, and a clear proposal on how funds will be used and accountability for their use.

The Beaufort Sea Panel re-emphasized long-term, formalized funding for intervenors by suggesting: "Intervenor funding be made available for all future EARP reviews and that funding be restricted to those participants who would be significantly affected by the proposal under review."

A study group, established by FEARO in 1987 to review EARP procedures, raised two additional issues related to intervenor funding. First, should the project proponent or government pay for participant funding? Second, who should allocate the funds? The study group urged strongly that panels have no role in distributing funds, so as to avoid a perception of bias. The group noted the useful precedent established in the EIA of military flying activities in Labrador and Quebec. A special funding administration committee set criteria for funding eligibility and distributed funds. The study group also suggested a permanent funding committee to develop ongoing expertise (FEARO 1988b).

Increasing Public Information

While dissemination of information to the public may be adequate at the public review stage through a required pre-hearing public information program (section 28[1] of the Guidelines Order) and through required public access to all information submitted to a panel (section 29[2]), the federal EARP does not ensure full disclosure of information following project approval. Panel recommendations become binding only by ministerial discretion. No requirement exists for ministers to provide reasons for accepting or rejecting panel recommendations, and public information is not required as to progress in implementing recommendations.

The Beaufort Sea Panel perhaps has pointed a direction for future panels, by recognizing this weakness. It suggested (recommendation 83)

that the Department of Indian Affairs and Northern Development publish a yearly report describing implementation of recommendations and
giving reasons for not accepting particular recommendations.

The National Consultation Workshop of March 1988 emphasized
regular post-project audits as an additional educational tool. An assessment
of project consequences would not only identify environmental problems needing further resolution but also increase understanding of
impact prediction and mitigation (FEARO 1988a). Post-assessment
informational requirements should be legally enshrined to assure full
disclosure and consistency.

Adding Certainty

While the procedures of past federal EARP panels have been rather ad
hoc, fluctuating from panel to panel according to the chair's view of
proper procedure (Emond 1983), there has been a movement toward
greater certainty. The EARP Guidelines Order requires a panel to allow
the public sufficient time to examine and comment on information
submitted prior to a public hearing (section 29[2]) and requires FEARO
to provide written procedures and provide advice to panels so as to
attain procedural consistency (section 35[d]). In 1985 FEARO issued core
procedures and rules for public meetings which establish general rules
such as publishing procedures in advance of public hearings and making
reports of technical specialists available to the public (FEARO 1985). The
Beaufort Sea Panel developed comprehensive procedures, including requiring presentations of expert opinion or technical fact to be distributed
to other intervenors at least one week before scheduled presentation (in
its final report, the panel recommended two weeks).

Since EARP usually involves not just factual disputes but also conflicts
over values and interests, the choice of informal procedures over formal
quasi-judicial procedures seems justified. As noted by the study group,
"it is ... very difficult to imagine how it would be possible to cross-
examine an intervenor on his or her values and choices." Formal procedures may also lessen the time available for public contribution and
may substantially increase the costs of hearings (FEARO 1988b).

Although numerous procedural changes are possible, including giving panels statutory power to subpoena witnesses or documents or to
require participants to answer questions (Emond 1983), it would be
useful to establish more certain time frames, in fairness particularly to
proponents who, for financial reasons, may be concerned with long
delays and efficiency. There might be, for example, a minimum of 60
days' public notice required before panel hearings, 45 days for public

notice and comment on a completed EIS, a maximum of 65 days for a panel to decide on the sufficiency of an EIS, and 45 days for a panel to submit a final report after the close of hearings.

Clearer Criteria

The federal EARP operates without clear criteria for determining "significance" of environmental effects (the trigger point for formal EARP review) or for "acceptability," which could be a standard for a recommendation to proceed or not proceed. The Guidelines Order states that all proposals that might significantly affect the environment must be referred to public review (section 3) but gives no definitions or criteria for determining significance. Section 12(e) appears to leave determination of significance to negotiation between FEARO and initiating departments. Panels are given a free hand in making final recommendations and are given no pre-established policy guidance concerning how to assess acceptability.

At least two reasons exist for establishing firm substantive criteria in legislation or by cabinet order. In a democratic society, elected officials have a duty to address fundamental policy questions (Emond 1978) and should, therefore, provide clear policy direction for appointed administrators to follow. The principle of fairness dictates that all players in the assessment game know all the rules beforehand, including when the game gets played (significance) and how to attain a winning score (acceptability).

Numerous approaches to establishing criteria are possible. Significance might be defined in terms of size (for example, any road over 1 km in length and over 10 m wide), type (for example, any hydroelectric project), geographical area affected, dollar value, and ecological effects, such as potential to affect an endangered species or an area of special ecological significance. Significance might also be defined in terms of concern by a predesignated number of citizens. Acceptability might also be determined according to various standards, including cost-benefit (overall project benefits must outweigh project costs), ecological (no substantial loss of critical habitat to an endangered or rare species), technological (use of best available, best practicable, or best feasible technology), and human acceptability (e.g. to a majority of the potentially affected community).

Emond summarized the role of objective criteria: "Ideally ... the government should state its policies in the legislation or procedures, and let the details of the policies evolve through a specialized board in the context of individual assessments" (Emond 1978).

Ensuring Public Review

Section 13 of the EARP Guidelines Order leaves some ministerial discretion about public environmental assessment hearings: "If public concern about the proposal is such that a public review is desirable, the
initiating department shall refer the proposal to the Minister for public
review by a Panel." In contrast, the Ontario Environmental Assessment
Act provides a model of more certainty and citizen power in the initiation of public hearings: an individual may write to the minister of
environment to demand a public hearing, which may be denied only
if the minister determines "that the requirement is frivolous or vexatious
or that a hearing is unnecessary or may cause undue delay" (section
12[2]).

Ensuring Judicial Review

The ability of a citizen to challenge the EARP process before a court –
for example, to force a public review when one has not been held, to
challenge the sufficiency of scientific data, or to contest procedural
fairness – remains uncertain. The legal effect of the 1984 Guidelines
Order remains questionable, since the term *guidelines* is used, even though
many of the provisions seem to impose mandatory obligations (Elder
1985). Courts have been hesitant to impose procedural reins on policy-
making bodies (Mullan 1983), such as EARP, which tend to deal with
broad issues of social, economic, or environmental policy rather than
individual rights. Some courts continue to take a conservative attitude
on standing (Lucas and McCallum 1975; Mullan 1983).

Although the Task Force on Environmental Impact Policy and Procedure in 1972 recommended a legislative base and judicial review for
EARP, the federal government has not taken such steps, perhaps largely
because it feared a quagmire of adversarial litigation, as in the United
States. Jack Davis, then minister of the environment, stated: "We will
not follow the highly legalistic approach developed in the United States
[under the National Environmental Policy Act, 1969] and found want-
ing. Canada is striking out on its own. We are more flexible" (quoted
in Emond 1978).

A legislative base for EARP could bestow standing to all interested
persons and clarify the bases for judicial intervention. Judges, removed
from political lobbying and electoral pressures, may play an important
"watch-dog" function in assuring that all public voices are heard and
in guaranteeing the rules of the assessment game. Administrators might
become more sensitive and careful about procedural fairness and construction of an adequate informational record. The public would be
granted an additional avenue of participation.

Additional reasons exist for legislating the process. EARP legislation could implement subpoena powers, establish FEARO as an independent agency reporting to Parliament through the minister of the environment, and grant legislative authority for considering socio-economic issues. Legislation might also force federal officials to regard the process more seriously (FEARO 1988b).

NEW DIRECTIONS IN OFFSHORE MANAGEMENT

Arrangements for offshore legal jurisdiction and management are not etched in stone but are open to change based on political factors, as demonstrated by legal history in both the United States and Canada.

For example, in 1947 the US Supreme Court granted the federal government the right to exploit the petroleum and mineral resources of the seabed off California, even though the state had granted leases for offshore exploration since the early 1920s. The court based federal jurisdiction on three legal pillars – an earlier international claim by federal officials to a three-mile territorial sea, national defence and security, and protection of foreign commerce and international relations. However, after political pressure exerted by coastal states, Congress in 1953 passed the Submerged Lands Act, which granted coastal states the right to exploit and regulate offshore resources out to three nautical miles, subject to federal powers over "commerce, navigation, national defense and international affairs."

In 1984 the Supreme Court of Canada ruled that the federal government had the sovereign right to exploit the seabed resources beyond the Newfoundland territorial sea. However, in subsequent federal-provincial negotiations, Canada and Newfoundland agreed to the Atlantic Accord, which established revenue-sharing and a joint board to oversee offshore oil and gas activities.

Four evolving processes, in particular, promise to push arctic offshore management in new directions, toward greater participation by northern residents and long-term planning. A land use planning process for Northwest Territories, agreed to in July 1983 by four aboriginal organizations, DIAND, and the government of Northwest Territories (GNWT), is being implemented (Donihee 1985). An NWT Land Use Commission, with equal governmental and native representation, has been granted a planning function for the arctic offshore as well as land, and regional commissions have been established for the Beaufort Sea–Mackenzie Delta and Lancaster Sound regions. As currently formulated, the process appears to be primarily advisory, consultative, and consensus-seeking. The land use planning process leaves decision-making authority with the ministers of DIAND and NWT Renewable Resources, who would have to justify decisions contrary to land use plans. Federal jurisdictional authority over the offshore remains unchanged.

A second process is the implementation of the 1984 COPE Agreement, which settled Inuvialuit land claims in the western Arctic and established new policy principles and participatory rights for the Inuvialuit. The agreement establishes three clear policy principles – preservation of Inuvialuit cultural identity, Inuvialuit participation in northern and national economy and society, and protection of arctic wildlife and environment (section 1).

To implement the principles, the agreement provides for various planning and advisory mechanisms. Wildlife management advisory councils, one for Yukon North Slope and one for Northwest Territories, are to draw up wildlife management plans and recommend harvest quotas (sections 12[8] and 14[6]). Two land use planning commissions being established, for Yukon and for Northwest Territories, are to have sub-groups focusing on the Beaufort Sea Region (section 7[17]). Marine-related development projects that might harm wildlife harvesting may be subject to a joint Inuvialuit-government environmental impact screening and review process (sections 11[1], 12[5][B][ii], 13[21]). Development proposals for Yukon North Slope, a special conservation area, that may harm wildlife, habitat, or wildlife harvesting must be subjected to a public environmental impact assessment and review process (section 12[1]). Other advisory mechanisms include an Inuvialuit Game Council (section 14[8]), a Fisheries Joint Management Committee (section 14[7]), and a Research Advisory Council (section 14[11]). In addition, a Yukon North Slope Annual Conference should promote management co-ordination among native persons, governments, and the private sector (section 12[a]). Questions concerning interpretation, implementation, or application of the final agreement may be referred to an arbitration board, consisting of representatives from federal and territorial governments, as well as native and industry representatives. An award of the board would be subject to review by the Federal Court of Appeal (section 18).

A third process is ongoing negotiation of land claims for the central and eastern Arctic, which could also lead to new institutional mechanisms and perhaps a realignment of political powers over the arctic offshore. The Tungavik Federation of Nunavut (TFN), the negotiating arm of Inuit in the central and eastern Arctic, has negotiated a series of agreements in principle, which provides for new agencies, such as a Nunavut Planning Commission, a Nunavut Impact Review Board, a Nunavut Water Board, and a Nunavut Wildlife Management Board. A task force appointed by Minister of DIAND David Crombie strongly supported inclusion of Inuit in resource-sharing and shared management arrangements (see Task Force to Review Comprehensive Claims Policy 1985).

A fourth process is the negotiation of northern accords respecting oil and gas resource management and revenues. In September 1988 the federal government signed agreements in principle with the northern governments, promising to manage jointly offshore hydrocarbon developments and to share oil and gas revenues.

The evolving processes applicable to the offshore raise numerous questions. Will northern land use planning eventually receive a legislative base (Task Force on Northern Conservation 1984)? How will land use planning relate to other federal decision-making or advisory processes, such as the shipping control authority responsible for designating and studying arctic shipping routes or the approval and site-selection process for ocean dumping? What will be the offshore boundaries of regional land use plans? The 18 June 1984 Letter of Agreement on Northern Land Use Planning between the minister of renewable resources (GNWT) and the minister of DIAND defined only generally areas of applicability: "The plans will provide for the conservation, development and utilization of land, resources, inland waters and the offshore." A Land Use Planning Policy Advisory Committee will identify specific planning regions. Numerous offshore "land" use planning boundaries are possible, including the seaward extent of the territorial sea (12 nautical miles), the outer edge of the landfast ice, out to 100 nautical miles from shore (the present coverage of the Arctic Waters Pollution Prevention Act), out to 200 nautical miles (the extent of Canadian fisheries zones), or out to the edge of the continental margin (which may extend beyond 200 nautical miles). Given strong reliance by coastal communities on marine living resources and the tendency for offshore ecosystem relationships to ignore human political and legal boundaries, the planning process should perhaps apply to the maximum offshore area possible for national jurisdiction, namely out to 200 nautical miles or the edge of the continental margin, whichever is greater.

The Lancaster Sound Planning Region has been defined in a second draft plan not to cover the maximum extent of national offshore jurisdiction. The regional land use plan will apply to approximately 1.5 million sq km, with the northern limit being the limit of Canada's fishing zone. (The first draft referred to the jurisdictional limit allowed under international law.)

Will land use planning remain advisory in nature, or will the process eventually gain additional "legal teeth"? No clear procedure exists for resolving potential federal-territorial conflict. A possible approach would be to adopt a consistency standard, along with mediation and arbitration mechanisms, to require review of alleged inconsistencies between land use plans and federal-related offshore activities.

The first draft of the Lancaster Sound plan recommended a contact person in each community to receive non-compliance complaints and establishment of the Northwest Territories Land Use Planning Commission as the preferred resolution mechanism. However, the plan noted the potential need for further mechanisms: "If the plan is not being followed, conflict resolution mechanisms will be required" (Lancaster Sound Regional Land Use Planning Commission 1987).

There are other questions as well. What is the future of EARP? Will the process be supplanted through land claims by new offshore impact boards, or will EARP become a supplemental process? Will the territorial governments eventually develop their own independent environmental assessment processes? How will the federal EARP evolve? Recommendations on EARP reform were submitted to the federal cabinet in the spring of 1988, but as of September 1988 no changes had been announced. How will future constitutional and political development in the north affect offshore management? For the time being, at least, more questions remain than answers.

CONCLUSION

The challenges facing industrial proponents wishing to develop shipping in the Arctic are monumental and numerous. Ethical reality – the human condition of conflicting interests and values – not only explains fragmented decision-making among competing departments and the fixation of all parties on legal structure and power, but also calls for balanced and consensual development. Major commercial shipping in the Arctic should occur only through phased, small-scale development and with the consent of northern residents. Realities of risk assessment – differing perceptions of risk by experts and the public – call not only for full public participation in expressing perceptions but also for development of public benefits. Northern residents are much more likely to accept risks of development if social and economic benefits are conferred through such vehicles as training, employment, and revenue-sharing.

Numerous scientific issues roughen the waters of environmental assessment. Long-term scientific research programs are essential, but funds seem to be ever dwindling. Temporal and spatial boundaries of scientific studies need to be expanded, while the public needs to help focus studies on valued ecosystem components (Beanlands and Duinker 1983). Environmental assessment must be viewed as a continuing process of "learning by doing" and requires the support of a long-range planning process.

Procedural and substantive strengthenings to the assessment process are called for on numerous fronts. There is a need to assess not just individual projects but also the cumulative effects of many projects and the environmental effects of government programs and policies. There should be a move from self-assessment by self-interested departments to independent assessment by FEARO. Public involvement should be strengthened through sure funding of intervenors, innovative negotiation or mediation processes, and participation in post-assessment monitoring. Governments should have to explain acceptance or rejection of panel recommendations, and post-assessment reports on implementation should be mandatory. Time frames should be clarified, and substantive criteria established for determining "significance" and "acceptability." The public should be granted clear access to judicial review, to assure fair application of the assessment process.

New directions in Arctic Ocean management seem to be emerging through land use planning and land claims that raise numerous unanswered questions and could also add to the management confusion. One thing is certain. Challenges to arctic shipping will continue, and changes in offshore management will evolve. Disillusionment and dissatisfaction over the managerial regime will probably not disappear but may gradually lessen through political devolution to the regional and community level. All of this makes us still travellers, and still on the road to Kingdom Come.

POSTSCRIPT

Further strengthening of the federal EARP was promised by the minister of the environment in an announcement on 20 October 1988. Key reforms to be implemented through new legislation would include: mandatory environmental assessment for all activities within federal decision-making authority; FEARO being given power to determine the need for public review, with appeal of a FEARO decision possible to the environment minister; subpoena powers for EARP panels; overseas aid projects being subjected to environmental assessments; and use of mediators to resolve environmental disputes.

The legal status of the federal EARP Guidelines Order was clarified in a recent decision by the Federal Court of Canada, Trial Division, in *Canadian Wildlife Federation Inc. v. Minister of the Environment* (Court No. T-80-89, 10 April 1989). The court upheld the Canadian Wildlife Federation's argument that the EARP Guidelines Order is a legal enactment or regulation, not just a discretionary policy. Therefore, before issuing a licence, under the International River Improvements Act, that would allow construction of the Rafferty and Alameda dams in Saskatchewan, the minister of the environment was duty-bound to assure that a federal environment assessment occurred, including a review of potential effects of the projects on North Dakota and Manitoba. The case, subject to appeal, clarifies the mandatory nature of federal environmental assessment review and gives the public an additional legal lever for pressuring government officials to follow the provisions of the Guidelines Order.

Bibliography

ABBREVIATIONS

APP   Arctic Pilot Project
CARC   Canadian Arctic Resources Committee
CWS   Canadian Wildlife Service
DFO   Department of Fisheries and Oceans (Canada)
DIAND Department of Indian Affairs and Northern
       Development (Canada)
POAC   *Ports and Offshore Engineering under Arctic
       Conditions*, biennial publication of conference
       proceedings

Abels, G. 1892. "Measurement of the Snow Density at Ekaterinburg during the Winter of 1890–1891." *Academia Nauk, Memoirs.* 669

Acoustical Society of America. 1980. *Proceedings, Workshop on the Interaction between Man-made Noise and Vibration and Arctic Marine Wildlife.* San Diego

Agerton, D.J., and Kreider, J.R. 1979. "Correlation of Storms and Major Ice Movements in the Nearshore Alaskan Beaufort Sea." POAC 1:177–9

Ainley, D.G., Grau, C.R., Roudybush, T.E., Morrell, S.H., and Utts, J.M. 1981. "Petroleum Ingestion Reduces Reproduction in Cassin's Auklets." *Marine Pollution Bulletin* 12: 314–17

Albers, P.H. 1977. "Effects of External Applications of Fuel Oil on Hatchability of Mallard Eggs." In D.A. Wolfe, (ed.). *Fate and Effects of Petroleum Hydrocarbons in Marine Ecosystems and Organisms: Proceedings of a Symposium, Seattle, Washington, 1976.* New York: Pergamon. Pp. 158–63

Albery, Pulleritts, Dickson and Associates. 1978. *A Study of Marine Transportation of Oil and LNG from Arctic Islands to Southern Markets.* Ottawa: Transport Canada

Alexander, V., and Niebauer, H.J. 1981. "Oceanography of the Eastern Bering Sea Ice-edge in Spring." *Limnology and Oceanography* 26:1111–25

Andreas, E.L., Tucker, W.B., and Ackley, S.F. 1984. "Variation of the Drag Coefficient across the Antarctic Marginal Ice Zone." MIZEX Bulletin 3. *Modelling the Marginal Ice Zone.* CRREL Special Report 84–7:63–71

Andrews, R.N.L. 1981. "Value Analysis in Environmental Policy." In D.E. Mann (ed.). *Environmental Policy Formation.* Lexington: Lexington Books. Pp. 137–47

Anjum, S. 1985. "Land-use Planning in the North Slope Borough." In *National and Regional Interests in the North." Proceedings, Third National Workshop on People, Resources and the Environment North of 60 degrees.* Ottawa: CARC

Anonymous 1970–3. *Report of the Task Force – Operation Oil. (Clean-up of the Arrow Spill in Chedabucto Bay).* Vols. I–IV. Ottawa: Ministry of Transport

APP (Arctic Pilot Project). 1981a. *Arctic Pilot Project Integrated Route Analysis.* Calgary: Petro-Canada. 522 pp.

– 1981b. *The Question of Sound from Icebreaker Operations. Proceedings, Workshop Sponsored by Petro-Canada.* Calgary. 350 pp.

– 1982. *Application to the National Energy Board,* Volume III, Part D, *Marine Transportation.* Exhibit #30. Ottawa

"Arctic Tankers through the Northwest Passage." 1982. *Beaufort* (December):21–7

" 'Argo Merchant' Disaster 'Inevitable.' " 1977. *Marine Pollution Bulletin* 8:124

Ashmole, N.P. 1971. "Seabird Ecology and the Marine Environment. In D.S. Farner and J.R. King (eds.). *Avian Biology,* Vol. I. New York: Academic Press. Pp. 223–86

Austin-Smith, P.J. 1968. "Late Winter Oil Pollution in the Bay of Fundy, Nova Scotia." *Canadian Field-Naturalist* 82:145–6

Bailey, N.T.J. 1959. *Statistical Methods in Biology.* London: English Universities Press. 200 pp.

Bankes, N., and Thompson, A.R. 1980. *Monitoring for Impact Assessment and Management: An Analysis of the Legal and Administrative Framework.* Vancouver: Westwater Research Centre. 80 pp.

Barnes, J.C., and Bowley, C.J. 1974. *The Application of ERTS Imagery to Monitoring Sea Ice.* ERT Document 0408–5F, (US) National Aeronautics and Space Administration. 95 pp.

Barr, W. 1985. *The Expeditions of the First International Polar Year, 1882–83.* Arctic Institute of North America Technical Paper No. 29. Calgary: University of Calgary. 222 pp.

Beanlands, G.E., and Duinker, P.N. 1983. *An Ecological Framework for Environmental Impact Assessment in Canada.* Halifax: Institute for Resource and Environmental Studies, Dalhousie University, and Federal Environmental Assessment Review Office

"Beaufort Sea Project in Retrospect." 1982. *APOA Review* (spring/summer):21–7

BEMP (Beaufort Environmental Monitoring Project). 1985. *Final Report*. Ottawa: ESL Ltd., LGL Ltd., ESSA Ltd., Arctic Sciences Ltd., and Arctic Laboratories Ltd., for Environment Canada and DIAND

Bennett, E.B. 1978. Water Cooling versus Thermal Ice Management. Unpublished manuscript

Berger, T.R. 1984. "Development from the Perspective of Human Ecology." *Environments* 16(3):2–9

Bilello, M.A. 1961. "Formation, Growth and Decay of Sea-ice." *Arctic* 14(1):3–24

Birch, J.R., Fissel, D.B., Lemon, D.D., Cornford, A.B., Herlinveaux, R.H., Lake, R.A., and Smiley, B.D. 1983. *Arctic Data Compilation and Appraisal*. Vol. v. *Baffin Bay: Physical Oceanography – Temperature, Salinity, Current and Water Levels*. Canadian Data Report of Hydrography and Ocean Sciences No. 5, Department of Fisheries and Oceans (Canada), Sidney, British Columbia, 372 pp.

Birch, J.R., Fissel, D.B., Lemon, D.D., Cornford, A.B., Lake, R.A., Smiley, B.D., MacDonald, R.W., and Herlinveaux, R.H. 1983. *Arctic Data Compilation and Appraisal*. Vol. I. *Northwest Passage: Physical Oceanography – Temperature, Salinity, Currents and Water Levels*. Canadian Data Report of Hydrography and Ocean Sciences No. 5, Department of Fisheries and Oceans (Canada), Sidney, B.C. 262 pp.

Birkhead, T.R., Lloyd, C., and Corkhill, P. 1973. "Oiled Birds Successfully Cleaning Their Plumage." *British Birds*. 66:535–7

Boothroyd, P., and Rees, W. 1981. *Public Policy and the Northwest Passage*. Ottawa: CARC. 87 pp.

Bourne, W.R.P. 1974. "Guillemots with Damaged Primary Feathers." *Marine Pollution Bulletin* 5:88–90

– 1976. "Seabirds and Pollution." In R. Johnson (ed.). *Marine Pollution*. London: Academic Press. 403–501

– 1977. "Ecofiasco." *Marine Pollution Bulletin* 8:121–2

– 1980. "*Scenic* in Shetland." *Marine Pollution Bulletin* 11:58

Bourne, W.R.P., Parrack, J.D., and Potts, G.R. 1967. "Birds Killed in the 'Torrey Canyon' Disaster." *Nature* (London) 215:1123–5

Bradstreet, M.S.W. 1980. "Thick-billed Murres and Black Guillemots in the Barrow Strait Area, N.W.T., during Spring: Diets and Food Availability along Ice Edges." *Canadian Journal of Zoology* 58:2120–40

– 1982a. "Occurrence, Habitat Use and Behavior of Seabirds, Marine Mammals, and Arctic Cod at the Pond Inlet Ice Edge." *Arctic* 35:28–40

– 1982b. "Pelagic Feeding Ecology of Dovekies, *Alle alle*, in Lancaster Sound and Western Baffin Bay." *Arctic* 35:126–40

Bradstreet, M.S.W., and Cross, W.E. 1982. "Trophic Relationships at High
   Arctic Ice Edges." *Arctic* 35:1–12
Brosted, J., and Faegteborg, M. (eds.). 1985. *Native Power: The Quest for Au-
   tonomy and Nationhood in Indigenous Peoples.* Oslo: Universitetsforlaget
Brown, R.G.B. 1973. *Seabirds and Oil Pollution: The Investigation of an Offshore
   Oil Slick.* cws Progress Note No. 31. 4 pp.
– 1985. *Revised Atlas of Eastern Canadian Seabirds.* Ottawa: cws. 111 pp.
Brown, R.G.B. 1982. "Birds, Oil and the Canadian Environment." In J.B.
   Sprague, J.H. Vandermeulen, and P.G. Wells (eds.). *Oil and Dispersants in
   Canadian Seas – Research Appraisal and Recommendations.* Environmental Pro-
   tection Service Report eps 3-ec-82-2, Ottawa. Pp. 105–12
Brown, R.G.B., and Nettleship, D.N. 1983. "The Seabirds of Northeastern
   North America: Their Present Status and Conservation Requirements." In
   Proceedings of the Seabird Workshop, 3–5 August 1982, Cambridge, U.K. Cam-
   bridge: International Council for Bird Preservation
Brown R.G.B., Gillespie, D.I., Lock, A.R., Pearce, P.A., and Watson, G.H.
   1973. "Bird Mortality from Oil Slicks off Eastern Canada, February–April
   1970." *Canadian Field-Naturalist* 87:225–34
Brown, R.G.B., and Johnson, B.C. 1980. "The Effects of 'Kurdistan' Oil on
   Seabirds." In J. Vandermeulen (ed.). *Scientific Studies during the 'Kurdistan'
   Tanker Incident: Proceedings of a Workshop.* Bedford Institute of Oceanography
   Report Series No. bi-r-80-3, Dartmouth, ns. Pp. 203–11
Brown, R.G.B., and Nettleship, D.N. 1984a. "Capelin and Seabirds in the
   Northwest Atlantic." In D.N. Nettleship, G.A. Sanger, and P.F. Springer
   (eds.). *Marine Birds: Their Feeding Ecology and Commercial Fisheries Relation-
   ships. Proceedings of the Pacific Seabird Group Symposium, 6–8 January 1982,
   Seattle.* cws Special Publication. Pp. 184–94
Brown, R.G.B., and Nettleship, D.N. 1984b. "The Seabirds of Northeastern
   North America: Their Present Status and Conservation Requirements." In
   J.P. Croxall, P.G.H. Evans, and R.W. Schreiber (eds.). *Status and Conser-
   vation of the World's Seabirds.* International Council for Bird Preservation
   Technical Publication No. 2:85–100
Brown, R.G.B., Nettleship, D.N., Germain, P., Tull, C.E., and Davis, T.
   1975. *Atlas of Eastern Canadian Seabirds.* Ottawa: cws. 220 pp.
Bruchet, D., and Robertson, M. 1983. "Regulatory and Environmental Issues
   Associated with Marine Transportation." Arctic Plot Project paper delivered
   to Massachusetts Institute of Technology Science Conference. Calgary:
   Petro-Canada
Buckley, J.R., Gammelsrod, T., Johannessen, J.A., Johannessen, O.M., and
   Roed, L.P. 1979. "Upwelling: Oceanic Structure at the Edge of the Arctic
   Pack Ice in Winter." *Science* 203:165–67
Burstyn, H.L. 1968. "The Historian of Science and Oceanography." *Congrès
   d'histoire de l'oceanographie,* 1:665–75

Cairns, J., Jr. 1986. "Freshwater." In *Cumulative Environmental Effects: A Binational Perspective*. Ottawa: Supply and Services Canada. Pp. 39–43

Campbell, N.J. 1976. "A Historical Sketch of Physical Oceanography in Canada." *Journal of the Fisheries Research Board of Canada* 33 (pt. 4): 2155–67

Canada, Advisory Committee on Northern Development. 1973. *Science and the North: A Seminar on Guidelines for Scientific Activities in Northern Canada (15–18 October 1972, Mont Gabriel, Quebec)*. Ottawa. 287 pp.

Canadian Coast Guard. 1980. *Ice Operations – Arctic (Annual Report)*. Ottawa: Transport Canada

– 1983. *I.R.A. Review Group Report to the Environmental Advisory Committee on Arctic Marine Transportation*. Ottawa. 30 pp.

CARC (Canadian Arctic Resources Committee). 1980. *The Arctic Pilot Project: Submission to the Federal Environmental Assessment and Review Office*. Ottawa. 71 pp.

Carstens, T. 1977. "Maintaining an Ice-free Harbor by Pumping of Warm Water." POAC. 1:347–57

CCREM (Canadian Council of Resource and Environment Ministers). 1978. *Canadian Environmental Impact Assessment Processes: Discussion Paper*. Ottawa. 34 pp.

CEARC (Canadian Environmental Assessment Research Council). 1986. *Philosophy and Themes for Research*. Ottawa: Supply and Services

– 1988. *The Assessment of Cumulative Effects: A Resource Prospective*. Ottawa: Supply and Services

Clark, R.B. 1978. "Oiled Seabird Rescue and Conservation." *Journal of the Fisheries Research Board of Canada* 35:675–8

Coachman, L.K. 1969. "Physical Oceanography in the Arctic Ocean." *Arctic* 22(3):214–24

Collin, A.E., and Dunbar, M.J. 1964. "Physical Oceanography in Arctic Canada." *Oceanography and Marine Biology Annual Review* 2:45–75

Cooke, A. and Van Alstine, E. (eds.). 1984. *Sikumiut: The People Who Use the Sea Ice*. Ottawa: CARC. 153 pp.

Coon, N.C., Albers, P.H., and Szaro, R.C. 1979. "No. 2 Fuel Oil Decreases Embryonic Survival of Great Black-backed Gulls." *Bulletin of Environmental Contamination and Toxicology* 21:152–6

Cornford, A.B., Lemon, D.D., Fissel, D.B., Melling, H., Smiley, B.D., Herlinveaux, R.H., and MacDonald, R.W. 1982. *Arctic Data Compilation and Appraisal*. Vol. 1. *Beaufort Sea: Physical Oceanography – Temperature, Salinity, Currents and Water Levels*. Canadian Data Report of Hydrography and Ocean Sciences No. 5, DFO, Sidney, BC. 279 pp.

CPA (Canadian Petroleum Association). 1982. *Northern Oil and Gas Development – Opportunities and Responsibilities*. Proceedings of a Workshop, 25–28 May, Yellowknife, N.W.T. Calgary. 160 pp.

Cramp, S., Bourne, W.R.P., and Saunders, D. (eds.). 1974. *The Seabirds of Britain and Ireland*. London: Collins, 287 pp.

Cramp, S., and Simmons, K.E.L. (eds.) 1977. *Handbook of the Birds of Europe, the Middle East and North Africa*. Vol. I. *Ostrich to Ducks*. Oxford University Press. 722 pp.

Croxall, J.P. 1976. *The Effects of Oil on Seabirds*. International Council for the Exploration of the Sea, Publication CM 1976/G5. 6 pp.

Department of Transport. 1965. *Manice: Manual of Standard Procedures and Practices for Ice Reconnaisance*. Ottawa. 59 pp.

DFO (Department of Fisheries and Oceans). 1982. *The Implications of the Beaufort Sea Hydrocarbon Production Proposal to the Department of Fisheries and Oceans*. Ottawa

– 1984. *Position Statement on the Beaufort Sea Hydrocarbon Production Proposal*. ARCOD Working Paper 84–2, Winnipeg. 51 pp.

– 1985. *Response to the Final Report of the Environmental Assessment Panel on Beaufort Sea Hydrocarbon Production and Transportation*. ARCOD Working Paper 85–4, Winnipeg. 161 pp.

– 1986. *Directory of Marine and Freshwater Scientists in Canada*. Canadian Special Publication of Fisheries and Aquatic Sciences 73 (Rev.). 259 pp.

DIAND (Department of Indian Affairs and Northern Development). 1970. "Address by the Honourable J. Chretien to the House of Commons on Second Reading of the Arctic Waters Pollution Prevention Act Bill." Published in *Hansard*, April 16

– 1982. *Land Use Planning in Northern Canada (Draft)*. DIAND, Ottawa. 65 pp.

– 1983a. *The Beaufort Environmental Monitoring Program: Presentation at the Inuvik Technical Hearing (November)*. 4 pp.

– 1983b. *Requirements for Successful Land Use Planning: Presentation by Y. Dube, DIAND, to the Alaska Science Conference, 30 September, Whitehorse*

– 1983c. *The Wonders and Pitfalls of Land Use Planning in the Northern Regions of Canada: Presentation by Y. Dube, Director General, Northern Affairs Program, DIAND, to the Banff Centre School of Management, 8 March*. Ottawa

– 1984. *The Western Arctic Claim: The Inuvialuit Final Agreement*. Ottawa

Dickins, D.F., and Wetzel, V.F. 1981. "Multi-Year Pressure Ridge Study, Queen Elizabeth Islands." POAC 2:765–76

Dietrich, J., Zorn, R., and Hasle, A. 1979. "Iceberg Investigation along the West Coast of Greenland." POAC. 1:221–39

Dirschl, H.J. 1982. *The Lancaster Sound Region: 1980–2000: Issues and Options for the Use and Management of the Region*. Ottawa: DIAND. 101 pp.

Dohler, G. 1966. *Tides in Canadian Waters*. Ottawa: Canadian Hydrographic Service, Department of Mines and Technical Surveys. 16 pp.

Dome Petroleum Ltd. 1982. *Brief to Special Committee of the Senate on the Northern Pipeline*. Ottawa

Dome, Gulf, and Esso (Dome Petroleum Ltd., Gulf Oil Canada, and Esso

Resources Ltd.). 1983. *Closing Statement to the Beaufort EARP Panel.* Calgary. 5 pp.

Donihee, J. 1985. 'The Government of the Northwest Territories as a "New Regulator." ' In P. Seely (ed.). *Making the Frontiers Conventional: Proceedings of the Fourteenth Annual APOA/CPA-OOD Workshop.* Calgary: Pallister Resource Management Ltd. Pp. A-43–63

Dorcey, A.H.J. 1986. "Techniques for Joint Management of Natural Resources: Getting to Yes." In J.O. Saunders (ed.). *Managing Natural Resources in a Federal State.* Toronto: Carswell. Pp. 14–31

Dosman, E.J. 1979. "Streamlining Interdepartmental Coordination in the High Arctic." *Arctic Seas Bulletin* 1(1)

Douglas, M., and Wildavsky, A. 1982. *Risk and Culture.* Berkeley: University of California Press. 221 pp.

"Drilling Ban Being Urged in Lancaster Sound Area." 1984. *Calgary Herald* (May 10)

Drury, W.H. 1973. "Population Changes in New England Seabirds." *Bird-Banding* 44:267–73

Dunbar, M.J. 1951. *Eastern Arctic Waters.* Fisheries Research Board of Canada, Bulletin No. 88. 119 pp.

– 1979. "Keynote Address." *Marine Transportation and High Arctic Development (Symposium Proceedings, 21–23 March, Montebello, Quebec).* Ottawa: CARC. Pp. 1–11

– 1981. "The History of Oceanographic Research in the Waters of the Canadian Arctic Islands." In M. Zaslow (ed.). *A Century of Canada's Arctic Islands (Proceedings of the 23rd Symposium of the Royal Society of Canada, 11–13 August 1980, Yellowknife, N.W.T.).* Ottawa: Royal Society of Canada. Pp. 141–51

Duval, W.S., Martin, L.C., and Fink, R.P. 1981. *A Prospectus on the Biological Effects of Oil Spills in the Marine Environment.* Report prepared by ESL Environmental Science Ltd., Vancouver, for Dome Petroleum, Calgary. 92 pp. plus 2 appendices

Edmunds, S.W. 1981. "Environmental Policy: Bounded Rationality Applied to Unbounded Ecological Problems." In D.E. Mann (ed.). *Environmental Policy Formation.* Lexington: Lexington Books. Pp. 191–201

Elder, P.S. 1975. "An Overview of the Participatory Environment in Canada." In P.S. Elder (ed.). *Environmental Management and Public Participation.* Toronto: Canadian Environmental Law Research Foundation. Pp. 370–84

– 1985. "The Federal Environmental Assessment and Review Process Guidelines Order." *Resources* (Newsletter of the Canadian Institute of Resources Law) No. 13:4–5

Emond, D.P. 1978. *Environmental Assessment Law in Canada.* Toronto: Emond-Montgomery. 380 pp.

– 1983. "Fairness, Efficiency, and FEARO: An Analysis of EARP." In E.S. Case, P.Z.R. Finkle, and A.R. Lucas (eds.). *Fairness in Environmental and Social*

Impact Assessment Processes. Calgary: Canadian Institute of Resources Law. Pp. 49–74

– 1984. "Co-operation in Nature: A New Foundation for Environmental Law." *Osgoode Hall Law Journal* 22(2):323–48

Engelhardt, F.R. 1981. "Oil Pollution in Polar Bears: Exposure and Clinical Effects." In *Proceedings, Fourth Arctic Marine Oilspill Program Technical Seminar, Edmonton.* Ottawa: Environmental Protection Service.

Environment Canada. 1982. *Environment Canada's Proposed Response to Beaufort Sea Hydrocarbon Production.* Ottawa: Queen's Printer. 37 pp.

– 1984. *Sustainable Development: A Submission to the Royal Commission on the Economic Union and Development Prospects for Canada.* 18 pp.

Erckmann, W.J. 1986. "Commentary." In *Cumulative Environmental Effects: A Binational Perspective.* Ottawa: Supply and Services Canada. Pp. 19–21

ESL (Environmental Sciences Ltd.). 1982. *The Biological Effects of Hydrocarbon Exploration and Production Related Activities, Disturbances and Wastes on Marine Flora and Fauna of the Beaufort Sea Region.* Calgary: Dome Petroleum Ltd. 400 pp.

ESSA (Environmental and Social Systems Analysts Ltd.). 1982. *Review and Evaluation of Adaptive Environmental Assessment and Management.* Ottawa: Environment Canada. 116 pp.

Faulkner, G.N. 1985. "Comments on the Beaufort Environmental Assessment Review Panel." In P. Sealy (ed.). *Making the Frontiers Conventional: Proceedings of the Fourteenth Annual APOA/CPA-OOD Workshop.* Calgary: Pallister Resource Management Ltd. Pp. B-11–21

FEARO (Federal Environmental Assessment Review Office). 1979a. *Lancaster Sound Drilling: Report of the Environmental Assessment Panel.* Ottawa: Supply and Services Canada. 127 pp.

– 1979b. *Revised Guide to the Federal Environmental Assessment and Review Process.* Ottawa: Supply and Services. 12 pp.

– 1980. *Arctic Pilot Project (Northern Component): Report of the Environmental Assessment Panel.* Ottawa: Supply and Services Canada. 125 pp.

– 1984. *Beaufort Sea Hydrocarbon Production and Transportation: Final Report of the Environmental Assessment Panel.* Ottawa: Supply and Services Canada. 146 pp.

– 1985. *Environmental Assessment Panel: Procedures and Rules for Public Meetings.* Ottawa: Supply and Services Canada. 5 pp.

– 1988a. *National Consultation Workshop on Federal Environmental Assessment and Reform: Report on Proceedings.* Ottawa: Supply and Services

– 1988b. *Public Review: Neither Judicial Nor Political But an Essential Forum for the Future of the Environment.* Ottawa: Supply and Services

Fimreite, N. 1974. "Mercury Contamination of Aquatic Birds in Northwestern Ontario." *Journal of Wildlife Management* 38:120–31

Firmreite, N., Holsworth, W.M., Keith, J.A., Pearce, P.A., and Gruchy, I.M.

1971. "Mercury in Fish and Fish-Eating Birds near Sites of Industrial Contamination in Canada." *Canadian Field-Naturalist* 85:211–20

Finkle, P. 1983. "Public Consultation: A General Overview." *Future Planner* 1(1):5–8

Finley, K.J., Miller, G.W., Davis, R.A., and Green, C.R. 1984. *Responses of Narwhal* (Mondon monocerus) *and Belugas* (Delphinapterus leucas) *to Ice-breaking Ships in Lancaster Sound – 1983.* Prepared by LGL Ltd. for DIAND, Ottawa. 117 pp.

Fish, K. 1983. *Parliamentarians and Science.* Ottawa: Science Council of Canada

Fisher, J. 1952. *The Fulmar.* London: Collins. 496 pp.

Fisher, J., and Lockley, R.M. 1954. *Sea-Birds.* London: Collins. 320 pp.

Fisher, J., and Vevers, H.G. 1943–4. "The Breeding Distribution, History and Population of the North Atlantic Gannet (*Sula bassana*)." *Journal of Animal Ecology* 12:173–213, 13:49–62

"Fisheries May Scuttle Ocean Research Scientists Fear." 1985. (Toronto) *Globe and Mail* (26 July)

Fissel, D.B., Cuypers, L., Lemon, D.D., Birch, J.R., Cornford, A.B., Lake, R.A., Smiley, B.D., MacDonald, R.W., and Herlinveaux, R.H. 1983. *Arctic Data Compilation and Appraisal.* Vol VI. *Queen Elizabeth Islands: Physical Oceanography – Temperature, Salinity, Currents and Water Levels.* Canadian Data Report of Hydrography and Ocean Sciences Number 5, DFO, Sidney, BC. 214 pp.

Fissel, D.B., Knight, D.N., and Birch, J.R. 1984. *An Oceanographic Survey of the Canadian Arctic Archipelago, March 1982.* Canadian Contractor Report of Hydrography and Ocean Science No. 15, Sidney, BC. 415 pp.

Fissel, D.B., Lemon, D.D., and Birch, J.R. 1982. "Major Features of the Summer Near-Surface Circulation of Western Baffin Bay, 1978 and 1979." *Arctic* 35(1):180–200

Fissel, D.B., and Marko, J.R. 1978. *A Surface Current Study of Eastern Parry Channel, NWT, Summer 1977.* Contractor Report Series 78–4, Fisheries and Environment Canada, Sidney, BC. 66 pp.

Fletcher, H.F. 1977. "Toward a Relevant Science: Fisheries and Aquatic Scientific Resource Needs in Canada." *Journal of the Fisheries Research Board of Canada* 34(7):1046–74

Foster, M., and Marino, C. 1986. *The Polar Shelf: The Saga of Canada's Arctic Scientists.* Toronto: NRC Press. 128 pp.

"Frontier Hetherington's Bread and Butter." 1984. *Calgary Herald* (9 September).

Gade, H.G., Lake, R.A., Lewis, E.L., and Walker, E.R. 1974. "Oceanography of an Arctic Bay." *Deep-Sea Research* 21:547–71.

Gamble, D.J. 1986. "Crushing of Cultures: Western Applied Science in Northern Societies." *Arctic* 39(1):20–3

Gaston, A.J. 1980. *Population, Movements and Wintering Areas of Thick-Billed Murres* (Uria lomvia) *in Eastern Canada.* CWS Progress Notes No. 110. 10 pp.

- 1982. "Migration of Juvenile Thick-Billed Murres through Hudson Strait in 1980." *Canadian Field-Naturalist* 96:30–4

Gaston, A.J., and Nettleship, D.N. 1981. *The Thick-Billed Murres of Prince Leopold Island.* cws Monograph Series No. 6, Ottawa. 350 pp.

Geraci, J.R. and St Aubin, D.J. 1980. "Offshore Petroleum Resource Development and Marine Mammals: A Review and Research Recommendation." *Marine Fisheries Review* 42:1–12

Geraci, J.R., and Smith, T.G. 1976. Direct and Indirect Effects of Oil on Ringed Seals (*Phoca hispida*) of the Beaufort Sea." *Journal of the Fisheries Research Board of Canada* 39:1976–84

Gerwick, B.C. (ed.). 1985. *Arctic Ocean Engineering for the 21st Century: Proceedings of the First Spilhaus Symposium, 14–17 October, Williamsburg, VA.* Washington, DC: Marine Technology Society

Gibson, R.B. 1975. "The Value of Participation." In P.S. Elder (ed.). *Environmental Management and Public Participation.* Toronto: Canadian Environmental Law Research Foundation. Pp. 7–39

Giovando, L.F., and Herlinveaux, R.H. 1981. *A Discussion of Factors Influencing Dispersion of Pollutants in the Beaufort Sea.* Pacific Marine Science Report 81–4, Institute of Ocean Sciences, Sidney, BC.

Gould, P.J., Forsell, D.J., and Lensink, C.J. 1982. *Pelagic Distribution and Abundance of Seabirds in the Gulf of Alaska and Eastern Bering Sea.* us Department of the Interior, Fish and Wildlife Service, Biological Services Program Report FWS/OBS-82/48, Washington, DC. 294 pp.

Greisman, P., and Lake, R.A. 1978. *Current Observations in the Channels of the Canadian Arctic Archipelago Adjacent to Bathurst Island.* Pacific Marine Sciences Report 78–23, DFO, Sidney, BC 127 pp.

Griffiths, F. 1979. *A Northern Foreign Policy.* Wellesley Papers 7/1979. Toronto: Canadian Institute of International Affairs. 90 pp.

"Gulf Makes Progress in Beaufort." 1982. *APOA Review* 5(2):7

Hachey, H.B. 1949. "Canadian Interest in Arctic Oceanography." *Arctic* 2(1):28–35

- 1961. *Oceanography and Canadian Atlantic Waters.* Fisheries Research Board of Canada, Bulletin 134, Ottawa. 120 pp.

Haggkvist, K. 1981. "Combination of a Sinking Warm Water Discharge and Air Bubble Curtains for Ice Reducing Purposes." POAC 2:1104–12

Hanson, A.J. 1985. "Why Assessments Are Needed." In P. Seely (ed.). *Making the Frontiers Conventional: Proceedings of the Fourteenth Annual APOA/CPA – OOD Workshop.* Calgary: Pallister Resource Management Ltd. Pp. B-2–10

Hartung, R. 1965. "Some Effects of Oiling on Reproduction of Ducks." *Journal of Wildlife Management* 29:872–4

- 1967. "Energy Metabolism in Oil-Covered Ducks." *Journal of Wildlife Management* 31:798–804

Hattersley-Smith, G., and Serson, H. 1970. "Mass Balance of the Ward

Hunt Ice Rise and Ice Shelf: A Ten Year Record." *Journal of Glaciology* 9(56):247–52

Henry, R.F. 1984. *Flood Hazard Delineation at Tuktoyaktuk.* Canadian Contractor Report of Hydrography and Ocean Sciences, No. 19. 117 pp.

Henry, R.F., and Foreman, M.G.G. 1977. *Numerical Model Studies of Semidiurnal Tides in the Southern Beaufort Sea.* Pacific Marine Science Report 77–11, Department of Fisheries and the Environment, Sidney, BC. 71 pp.

Henry, R.F., and Heaps, N.S. 1976. "Storm Surges in the Southern Beaufort Sea." *Journal of the Fisheries Research Board of Canada* 33:2362–76

Herlinveaux, R.H., and de Lange Boom, B.R. 1975. *Physical Oceanography of the Southeastern Beaufort Sea.* Beaufort Sea Technical Report No. 18, Department of the Environment, Victoria, BC. 97 pp.

Herlinveaux, R.H., de Lange Boom, B.R., and Wilton, G.R. 1976. *Salinity, Temperature, Turbidity and Meteorological Observations in the Beaufort Sea: Summer 1974, Spring and Summer 1975.* Pacific Marine Sciences Report 76–26, Environment Canada, Victoria, BC. 244 pp.

Hibler, W.D. 1979. "A Dynamic Thermodynamic Sea Ice Model." *Journal of Physical Oceanography* 9:815–46

Hibler, W.D., and Bryan, K. 1984. "A Large-Scale Ice/Ocean Model for the Marginal Ice Zone." In MIZEX Bulletin 3: *Modeling the Marginal Ice Zone.* CCREL Special Report 84–7. Pp. 1–7

Hildebrand, L.P. 1980. *An Assessment of Chronic Oil Pollution in Atlantic Canada.* Environmental Protection Service Report EPS-3-AR-81-2, Environment Canada, Dartmouth, NS. 23 pp.

Hocking, D., and Anderson, M. 1980. The Exploration, Discovery and Initial Development of the Eastern High Arctic. Unpublished manuscript. Petro Canada, Calgary. 45 pp.

Hodgins, D.O. 1983. *A Review of Extreme Wave Conditions in the Beaufort Sea.* Marine Environmental Data Service, DFO, Ottawa. 160 pp.

Holman, M. 1981. "Concluding Remarks." In *Arctic Pilot Project Workshop, "The Question of Sound from Icebreaker Operations."* Calgary: Petro-Canada

Holmes, W.N., and Cronshaw, J. 1977. "Biological Effects of Petroleum on Marine Birds." In D.C. Malins (ed.). *Effects of Petroleum of Arctic and Subarctic Marine Environments and Organisms.* New York: Academic Press. Pp. 359–8

Hope-Jones, P., Howels, G., Rees, E.I.S., and Wilson, J. 1970. "Effect of 'Hamilton Trader' Oil on Birds in the Irish Sea in May 1970." *British Birds* 63:97–110

Hunt, C.D., and Lucas, A.R. 1980. *Environmental Regulation – Its Impact on Major Oil and Gas Projects: Oil Sands and Arctic.* Calgary: Canadian Institute of Resources Law, University of Calgary. 160 pp.

Hurlburt, H.E. 1984. "The Potential for Ocean Prediction and the Role of Altimeter Data." *Marine Geodesy* 8(1–4):17–66.

IERC (Interdepartmental Environmental Review Committee). 1987. *Minutes of Meeting 87–1 on Arctic Marine Transportation.* Ottawa. 6 pp.

Inuit Tapirisat of Canada, Baffin Region Inuit Association and Labrador Inuit Association. 1982. *Brief to the Special Committee of the Senate on the Northern Pipeline.* Ottawa. 34 pp.

Ito, H. 1981. *On the Mechanics of the Fast Ice in the North Water Area.* Zurcher Geographische Schriften Vol II. 93 pp.

Jacobs, P. 1981. *People, Resources and the Environment: Perspectives on the Use and Management of the Lancaster Sound Region: Public Review Phase.* Ottawa: DIAND

Jacobs, P., and Palluq, J. 1983. *The Lancaster Sound Regional Study, Public Review: Public Prospect.* Ottawa: DIAND

Jarrell, R.A., and Roos, A.E. (eds.). 1981. *Critical Issues in the History of Canadian Science, Technology and Medicine: Proceedings, Second Conference on the History of Canadian Science, Technology and Medicine, Kingston, Ontario.* Thornhill: HSTC Publications. 262 pp.

Joelsen, S. 1981. Statement made by Joelsen, member of Greenland Parliament in Fall of 1981 Home Rule Parliament Debate. Translated by, published by, and available from the Greenland Home Rule Information Service, P.O. Box 1020, DK-3900, Nuuk, Greenland

Joerg, L. 1928. *Problems of Polar Research.* New York: American Geographical Society. 479 pp.

Johnson, L. 1983. *Assessment of the Effects of Oil on Arctic Marine Fish and Marine Mammals.* Canadian Technical Report of Fisheries and Aquatic Sciences No. 1200, DFO, Winnipeg. 15 pp.

Johnson, R.A. 1940. "Present Range, Migration and Abundance of the Atlantic Murre in North America." *Bird-Banding* 11:1–17

Johnstone, K. 1977. *The Aquatic Explorers: A History of the Fisheries Research Board of Canada.* Toronto: University of Toronto Press. 342 pp.

Jull, P. 1985. "The Aboriginal Option: A Radical Critique of European Values." *Northern Perspectives* 13(2):10–12

– 1986. *Politics, Development and Conservation in the International North.* Ottawa: CARC. 107 pp.

Jull, P., and Bankes, N. 1984. "Inuit Interests in the Arctic Offshore." In *Proceedings of the Third National Workshop on People, Resources and the Environment North of 60 Degrees, Yellowknife.* Ottawa: CARC

Kampp, K. 1982. *Den Kortnaebbede lomvie* Uria lomvia *i Gronland – vandringer, mortalitet og beskydning: en analyse af 35 ars ringmaerkninger.* Kobenhavns Universitet, Specialerapport til Naturvidenskabelig Kandidateksamen. 148 pp. (Translated from Danish, 1983, by CWS, Dartmouth, NS)

Klausner, S.Z., and Foulks, E.A. 1982. *Eskimo Capitalists: Oil, Politics and Alcohol.* Totowa, NJ: Allenheld and Osmun

Kovacs, A. 1983. "Characteristics of Multi-Year Pressure Ridges." POAC 3:173–82

Kupetskiy, V.N. 1962. *The "North Water," A Permanent Polynya in Baffin Bay*. Moscow State Oceanographic Institute. Trudy. 70 pp. Translated by Moira Dunbar

Lachapelle, A. 1981. "Winds and Waves in Lancaster Sound." POAC 2:830–42

Lake, R.A. 1981. A Working Paper on the Physical Oceanography of the Northwest Passage. Unpublished manuscript. Institute of Ocean Sciences, Sidney, BC

Lancaster Sound (NWT) Regional Land Use Planning Commission. 1987. *The Lancaster Sound Regional Land Use Plan (First Draft)*. Pond Inlet, NWT

Lang, R. 1981. "The Review Side of EIA: Lessons from a Panel Member." In A. Armour (ed.). *Issues in Environmental Impact Assessment*. Downsview, Ontario: York University, Faculty of Environmental Studies. Pp. 88–104

Langford, J.W. 1976. "Marine Science, Technology, and the Arctic: Some Questions and Guidelines for the Federal Government." In E.J. Dosman (ed.). *The Arctic in Question*. Toronto: Oxford University Press. Pp. 163–92

Langleben, M.P. 1969. "Albedo and Degree of Puddling of a Melting Cover of Sea Ice." *Journal of Glaciology* 8(54):407–12

Larkin, P.A. 1984. "A Commentary on Environmental Impact Assessment for Large Projects Affecting Lakes and Streams." *Canadian Journal of Fisheries and Aquatic Sciences* 41:1121–37

Leavitt, E., Mercer, B., Krakowski, E., and Schubert, K. 1983. "Real-Time Ice Forecasting in Support of Winter Drilling Operations in the Beaufort Sea." POAC 3:307–16

LeBlanc, R. 1981. Letter to John Roberts, Minister of Environment, from the Minister of Fisheries and Oceans. Ottawa

Levy, E.M. 1980. "Oil Pollution and Seabirds: Atlantic Canada 1967–77 and Some Implications for Northern Environments." *Marine Pollution Bulletin* 11:51–6

Levy, E.M., and A. Walton 1976. "High Seas Oil Pollution: Particulate Petroleum Residues in the North Atlantic." *Journal of the Fisheries Research Board of Canada* 33:2781–91

Lewis, E.L. 1979. *Arctic Ocean Heat Budget*. University of Bergen Geophysical Institute SCOR Report 52

Lewis, H.F. 1924. "List of Birds Recorded from the Island of Anticosti, Quebec." *Canadian Field-Naturalist* 38:43–6, 72–5

– 1931. "Five Years" Progress in the Bird Sanctuaries of the North Shore of the Gulf of St. Lawrence." *Canadian Field-Naturalist* 45:73–8

LGL Ltd. Environmental Research Associates. 1986. *Proceedings, Workshop on Selected Environmental Impacts of Ship Traffic in Lancaster Sound and Northern Baffin Bay*. Ottawa: DIAND. 123 pp.

Lock, A.R., and Ross, R.K. 1973. "The Nesting of the Great Cormorant (*Phalacrocorax carbo*) and the Double-Crested Cormorant (*Phalacrocorax auritus*) in Nova Scotia in 1971." *Canadian Field-Naturalist* 87:43–9

Loubser, J. 1984. "Development Centered on Man: Some Relevant Concepts from Canada." *Environments* 16(3):57–75

Lucas, A.R. 1979. "Regulation of Marine Operations in the Far North." Paper presented at the Association for Canadian Studies Conference, Canada and the Sea, Vancouver, 11 July

Lucas, A.R., and McCallum, S.K. 1975. "Looking at Environmental Impact Assessment." In P.S. Elder (ed.). *Environmental Management and Public Participation.* Toronto: Canadian Environmental Law Research Foundation. Pp. 307–18

Lumsden, R.W. 1980. *LNG Transportation by Sea.* Petro-Canada Environmental Safety Seminar, Calgary

MacInnis, A. 1985. "How Canadian Are Arctic Waters?" *Maritime Industries* 1(1):16–19

McEwan, E.H., and Koelink, A.F.C. 1973. "The Heat Production of Oiled Mallard and Scaup." *Canadian Journal of Zoology* 51:27–31

McLaren, P.L. 1982. "Spring Migration and Habitat Use by Seabirds in Eastern Lancaster Sound and Western Baffin Bay." *Arctic* 35:88–111

McLaren, P.L., and Renaud, W.E. 1982. "Seabird Concentrations in Late Summer along the Coasts of Devon and Ellesmere Islands, N.W.T." *Arctic* 35:112–17

McNeill, M.R., de Lange Boom, B.R., and Ramsden, D. 1978. *Radar Tracking of Ice in the Griffith Island Area of Barrow Strait, NWT.* Contractor Report Series *78–2*, Fisheries and Environment Canada, Sidney, BC. 105 pp.

MacNeill, M.R. and Garrett, J.F. 1975. *Open Water Surface Currents in the Southern Beaufort Sea.* Beaufort Sea Report No. 17, Department of the Environment, Victoria, BC. 113 pp.

McPhee, M.G. 1979. "The Effect of the Oceanic Boundary Layer on the Mean Drift of Pack Ice: Application of a Simple Model." *Journal of Physical Oceanography* 9:388–400

McPhee, M.G. 1980. "Physical Oceanography of the Seasonal Sea Ice Zone." *Cold Regions Science and Technology* 2:93–118

McTiernan, T.J. 1983. "The Development of Science and Technology in the Canadian Arctic." In *Science, Technology and Arctic Hydrocarbon Exploration: The Beaufort Experience: Proceedings, 34th Arctic Science Conference, Whitehorse.* Pp. 3-7

"Major Concerns of Beaufort Sea Production." *1981. Beaufort* 10:10–13

Mansfield, A.W. 1983. *The Effects of Vessel Traffic in the Arctic on Marine Mammals and Recommendations for Future Research.* Canadian Technical Report of Fisheries and Aquatic Sciences No. 1186, Ottawa. 97 pp.

Markham, W.E. 1981. *Ice Atlas, Canadian Arctic Waterways.* Ottawa: Supply and Services Canada

Marko, J.R. 1977. *A Satellite-Based Study of Sea Ice Dynamics in the Central Canadian Arctic Archipelago.* Contractor Report Series 77–4. Fisheries and Environment Canada, Sidney, BC. 106 pp.

Marmorek, D. 1984. *Semi-Direct Method of Demonstrating Interactions between*

Sample Sizes, Error and Detectability of Environmental Changes. Prepared for
ESSA Ltd. for Workshop on Systematic Techniques to Analyze Uncertainties
in Determination of Significant Ecological Change. Ottawa: National Research Council of Canada. 13 pp.

Marshall, D.W.I. 1984. "Challenges of Integrating Oil Production in Sparsely
Populated Areas." Paper presented at International Conference on Oil and
the Environment

Marshall, D.W.I. and Scott, P. 1982. "Environmental and Social Impact Assessment of the Beaufort Sea Hydrocarbon Production Proposal." Paper
presented at First International Conference on Social Impact Assessment

Maykut, G.A., and Grenfell, T.C. 1975. "The Spectral Distribution of Light
beneath First-Year Sea Ice in the Arctic Ocean." *Limnology and Oceanography*
70(4):554–63

Maykut, G.A., and Untersteiner, N. 1969. *Numerical Prediction of the Thermodynamic Response of Arctic Sea Ice to Environmental Changes.* Memorandum
RM-6093-PR, R and Corp., Santa Monica, Calif. 173 pp.

Melling, H. 1983. *Oceanographic Features of the Beaufort Sea in Early Winter.*
Canadian Technical Report of Hydrography and Ocean Sciences, No. 20,
DFO, Sidney, BC. 131 pp.

– 1984. Technical Comments on the Final Report of the Environmental Assessment Panel, Beaufort Sea Hydrocarbon Production and Transportation.
Memorandum to B. Smiley, Institute of Ocean Sciences, Fisheries and Oceans
Canada, Sidney, BC

Melling, H., Lake, R.A., Topham, D.R., and Fissel, D.B. 1984. "Oceanic
Thermal Structure in the Western Canadian Arctic." *Continental Shelf Research*
3(3):233–58

Milbrath, L.W. 1981. "Environmental Values and Beliefs of the General Public
and Leaders in the United States, England, and Germany." In D.E. Mann
(ed.). *Environmental Policy Formation.* Lexington: Lexington Books. Pp. 43–61

Miller, D. 1980. *Bridport Ice Management. Presentation Papers, Arctic Pilot Project,
Part 10.* EARP technical hearings, Resolute Bay, NWT, 23–29 April

Miller, D.S., Peakall, D.B., and Kinter, W.B. 1978. "Ingestion of Crude Oil:
Sublethal Effects in Herring Gull Chicks." *Science* 199:315–17

Miller, G.W. and Davis, R.A. 1984. *Distribution and Movements of Narwhals and
Beluga Whales in Response to Ship Traffic at the Lancaster Sound Edge – 1984.*
Prepared by LGL Ltd. for DIAND, Ottawa. 34 pp.

Milne, A.R. n.d. *Oil, Ice and Climate Change: Beaufort Sea Project.* Sidney, BC:
DFO

Milne, A.R., and Smiley, B.D. 1978. *Offshore Drilling in Lancaster Sound: Possible
Environmental Hazards.* Patricia Bay, BC: Institute of Ocean Sciences. 95 pp.

Moller, H. 1981. "The Influence of Low Frequency and Infrasonic Noise on
Man." In Arctic Pilot Project Workshop, "The Question of Sound from
Icebreaker Operations." Calgary: Petro-Canada. Pp. 310–20

Montevecchi, W.A., and Porter, J.M. 1980. "Parental Investments by Seabirds

at the Breeding Area with Emphasis on Northern Gannets (*Morus bassanus*)."
In J. Burger, B.L. Olla, and H.E. Winn (eds.). *Behavior of Marine Animals.*
Vol. IV. *Marine Birds.* New York and London: Plenum. Pp. 323–447

Morley, C.G. 1975. "The Legal Framework for Public Participation in Canadian Water Management." In P.S. Elder (ed.). *Environmental Management and Public Participation.* Toronto: Canadian Environmental Law Research Foundation. Pp. 40–83

Muench, R.D. 1983. "Mesoscale Oceanographic Features Associated with the Central Bering Sea Ice Edge: February – March 1981." *Journal of Geophysical Research* 88(C):2715–22

Mullan, D.J. 1983. "The Developing Law of Procedural and Substantive Fairness." In E.S. Case, P.Z.R. Finkle, and A.R. Lucas (eds.). *Fairness in Environmental and Social Impact Assessment Processes.* Calgary: Canadian Institute of Resources Law. Pp. 15–28

Munn, R.E. (ed.). 1979. *Environmental Impact Assessment: Principles and Procedures.* 2nd ed. Chichester: John Wiley. 190 pp.

Munro, D.A. 1986. "Environmental Impact Assessment as an Element of Environmental Management." In *Cumulative Environmental Effects: A Binational Perspective.* Ottawa: Supply and Services Canada. Pp. 25–30

Munro, J. 1982. Speech delivered to Canadian Club, Toronto

NEB (National Energy Board). 1982a. Exhibit #177 of the Arctic Pilot Project Hearings, entitled Report of the Environmental Assessment Panel – LNG Receiving Terminal (Arctic Pilot Project) Melford Point, Nova Scotia. Ottawa

– 1982b. Exhibit #178 of the Arctic Pilot Project Hearings, entitled Bureau d'audiences publiques sur l'environnement / Report of an Inquiring and Public Hearing – Gros Cacouna LNG Terminal. Ottawa

Needler, G.T. 1980. "Oceanographic Variability in the Operating Environment." In *Proceedings of the Ninth Environmental Workshop on Offshore Hydrocarbon Development.* Calgary: Arctic Institute of North America. Pp. 185–206

Nettleship, D.N. 1974. "Seabird Colonies and Distributions around Devon Island and Vicinity." *Arctic* 27:95–103

– 1980. *A Guide to the Major Seabird Colonies of Eastern Canada: Identity, Distribution and Abundance.* CWS, Studies on Northern Seabirds, Manuscript Report No. 97, Dartmouth, NS. 130 pp.

– 1985. "Breeding of Arctic Seabirds in Unusual Ice Years: The Thick-Billed Murre *Uria lomvia* in 1978." *Bedford Institute of Oceanography, BIO Review"* 84: 35–8

Nettleship, D.N., and Gaston, A.J. 1978. *Patterns of Pelagic Distribution of Seabirds in Western Lancaster Sound and Barrow Strait, Northwest Territories, in August and September 1976.* CWS Occasional Paper No. 39, Ottawa 40 pp.

Nettleship, D.N., and Smith, P.A. 1975. *Ecological Sites in Northern Canada.* Canadian Committee for the International Biological Programme, Conservation Terrestrial – Panel 9, Ottawa. 330 pp.

Newman, P.C. 1963. *Renegade in Power: The Diefenbaker Years*. Toronto: McClelland and Stewart. 414 pp.

Nicholls, H.B. 1986. *Environmental Advisory Committee on Arctic Marine Transportation: Review of Activities, 1981 THROUGH 1988*. Canadian Technical Report of Fisheries and Aquatic Sciences No. 1486, DFO, Dartmouth, NS.

Nordenskjold, O., and Mecking, L. 1928. *The Geography of the Polar Regions*. American Geographical Society Special Publication No. 8.

Norris, K.S. 1981. "Marine Mammals of the Arctic, Their Sounds and Their Relation to Alterations in the Acoustic Environment by Man-Made Noise." In APP Workshop, "The Question of Sound from Icebreaker Operations," Petro-Canada, Calgary. Pp. 304–9

Northern Frontier, Northern Homeland. 1977. *Report of the MacKenzie Valley Pipeline Inquiry*. Vol. II. Ottawa. 268 pp.

Norton, P., and McDonald, J.W. 1986. *Compilation of 1985 Industrial Activities in the Canadian Beaufort Sea*. Ottawa: DIAND. 132 pp.

Norton, P., Smiley, B.D., and de March, L. In press. *Arctic Data Compilation and Appraisal*. Vol. X. *Beaufort Sea: Biological Oceanography – Whales, 1848 through 1983*. Canadian Data Report of Hydrography and Oceanographic Science No. 5 Vol. 10. 407 pp.

Nunavut Constitutional Forum. 1985. *Building Nunavut Today and Tomorrow*. Ottawa

Nutt, D.C. 1951. "The 'Blue Dolphin' Labrador Expeditions, 1949 and 1950." *Arctic* 4(1):3–11

Ohlendorf, H.M. Risebrough, R.W., and Vermeer, K. 1978. *Exposure of Marine Birds to Environmental Pollutants*. US Department of the Interior, Fish and Wildlife Service, Wildlife Research Report No. 9, Washington, DC. 40 pp.

"Oil Tanker to Move through North Passage." 1985. *New North* (8 February): 1–2

O'Riordan, T. 1986. "EIA: Dangers and Opportunities." *Environmental Impact Assessment Review* 6:3–6

O'Riordan, T., and Sewell, W.R.D. 1981. "From Project Appraisal to Policy Review." In T. O'Riordan and W.R.D. Sewell (eds.). *Project Appraisal and Policy Review*. Toronto: John Wiley. Pp. 1–28

Orr, C.D., and Ward, R.M.P. 1982. "The Fall Migration of Thick-Billed Murres near Southern Baffin Island and Northern Labrador." *Arctic* 35:531–6

Orr, C.D., Ward, R.M.P., Williams, N.W., and Brown, R.G.B. 1982. "Migration Patterns of Red and Northern Phalaropes in Southwest Davis Strait and in the Northern Labrador Sea." *Wilson Bulletin* 94:303–12

Pallister, J. 1983. "The Development of Science and Technology in the Canadian Arctic." In *Science, Technology and Arctic Hydrocarbon Exploration: The Beaufort Experience: Proceedings of the 34th Alaska Science Conference, Whitehorse*. Pp. 8–17

"Panarctic Tanker Ships First Oil," 1985. *Arctic Policy Review* 3:2

"Panarctic Wins Bent Horn Development Approval." 1985. *Oilweek* (11 February):3

Parker, N., and Alexander, J. 1983. *Weather, Ice and Sea Conditions Relative to Arctic Marine Transportation*. Canadian Technical Report of Hydrography and Ocean Sciences No. 26, DFO, Sidney, BC. 211 pp.

Patten, P.R., Samaras, W.F., and McIntyre, P.R. 1980. "Whales, Move Over!" *American Cetacean Society Whalewatcher* 4:13–15

Peakall, D.B. 1975. "Physiological Effects of Chlorinated Hydrocarbons on Avian Species." In R. Haque, and V.H. Freed (eds.). *Environmental Dynamics of Pesticides*. New York: Plenum. pp. 347–60

Peakall, D.B., Hallett, D., Miller, D.S., and Kinter, W.B. 1980. "Effects of Ingested Crude Oil on Black Guillemots: A Combined Field and Laboratory Study." *Ambio* 9:28–30

Peck, G.S. 1978. *Arctic Oceanographic Data Report 1977: Western Viscount Melville Sound*. Data Report Series 73–3, Fisheries and Environment Canada, Burlington, Ont. 150 pp.

Perrins, C.M., and Birkhead, T.R. 1983. *Avian Ecology*. Glasgow: Blackie. 221 pp.

Pessah, E., and Robertson, M.R. 1986. "Industry Remarks." In *Environmental Advisory Committee on Arctic Marine Transportation: Review of Activities, 1981 THROUGH 1985*. Canadian Technical Report on Fisheries and Aquatic Science 1486, Ottawa. 37 pp.

Pharand, D. 1984. *The Northwest Passage: Arctic Straits*. Martinus Nijhoff Publisher. 236 pp.

Piatt, J.F., Nettleship, D.N. and Threlfall, W. 1984. "Net-Mortality of Common Murres and Atlantic Puffins in Newfoundland, 1951–81." In D.N. Nettleship, G.A. Sanger, and P.F. Springer (eds.). *Marine Birds: Their Feeding Ecology and Commercial Fisheries Relationships: Proceedings of the Pacific Seabird Group Symposium, 6–8 January 1982, Seattle*. CWS Special Publication, Ottawa. Pp. 196–206

Pimlott, D.H., Brown, D., and Sam, K.P. 1976. *Oil under Ice*. Ottawa: CARC. 17 pp.

Powers, K.D., and Rumage, W.T. 1978. "Effect of the 'Argo Merchant' Oil Spill on Bird Populations off the New England Coast." In *In the Wake of the 'Argo Merchant.' Proceedings of a Conference and Workshop, January 1978*. Kingston, RI: University of Rhode Island, Center for Ocean Management Studies. Pp. 142–8

Rae, R.W. 1951. "Joint Weather Project." ARCTIC 4(1):18–26

Rankin, M.N. and Duffey, E.A.G. 1948. "A Study of the Bird Life in the North Atlantic." *British Birds* 41 (supplement): 1–42

Rawles, M.E. 1960. "The Integumentary System." In A.J. Marshall (ed.).

Biology and Comparative Physiology of Birds. Vol. 1. New York: Academic Press. Pp. 190–240

Reed, John C. 1969. "The Story of the Naval Arctic Research Laboratory." *Arctic* 22(3):177–83

Rees, W.E. 1984. "Environmental Assessment of Hydrocarbon Production from the Canadian Beaufort Sea." *Environmental Impact Assessment Review* 4(3–4):539–55

Renaud, W.E., and McLaren, P.L. 1982. "Ivory Gull (*Pagophila eburnea*) Distribution in Late Summer and Autumn in Eastern Lancaster Sound and Western Baffin Bay." *Arctic* 35:141–8

Renaud, W.E., McLaren, P.L., and Johnson, S.R. 1982. "The Dovekie, *Alle alle*, as a Spring Migrant in Eastern Lancaster Sound and Western Baffin Bay." *Arctic* 35:118–25

Rice, A.L. 1975. "The Oceanography of John Ross" Arctic Expedition of 1818, a Re-appraisal." *Journal of the Society for the Bibliography of National History* 7(3):291–319

Richardson, N.H. 1987. "Northern Land Use Planning Agreements: A Commentary." In T. Fenge, and W.E. Rees (eds.). *Hinterland or Homeland? Land-Use Planning in Northern Canada.* Ottawa: CARC

Richardson, W.J., Green, G.R., Hickie, J.P., and Davis, R.A. 1983. *Effects of Offshore Petroleum Operations on Cold Water Marine Mammals: A Literature Review.* American Petroleum Institute Report No. 4370

Risebrough, R.W. 1969. "Chlorinated Hydrocarbons in Marine Ecosystems." In M.W. Miller, and G.G. Bery (eds.). *Chemical Fallout – Current Research on Persistent Pesticides.* Springfield, Ill.: Charles C. Thomas. Pp. 5–23

Roberts, B. (ed.). 1971. *The Arctic Ocean (Report of a Conference at Ditchley Park, 14–17 May).* Ditchley Park, UK: Ditchley Foundation. 49 pp.

Robertson, C.D. 1985. "Keeping Procedures Relevant – A Canadian Federal Perspective." Paper presented at the International Bar Association Seminar on Industry and the Regulatory Agencies, Stratford-upon-Avon, UK

Robinson, R. 1982. Notes for Address Given at the First International Conference on Social Impact Assessment, Ottawa

– 1985. "The Federal Role in Environmental Assessment." Paper presented at Canadian Institute of Resources Law Conference on Natural Resources Law, Ottawa

Roots, E.F. 1981. "Basic Science and Its Relation to Arctic Marine Engineering." POAC 3:1259–87

– 1982a. "Anniversaries of Arctic Investigation: Some Background and Consequences." *Royal Society of Canada Transactions* 373–90

– 1982b. "The Changing Marine Environment: Some Basic Considerations." In L. Rey (ed.). *The Arctic Ocean: The Hydrographic Environment and the Fate of Pollutants.* Monaco: Comité arctique international. Pp. 215–32

– 1984. "International and Regional Cooperation in Arctic Science: A Changing Situation." Paper presented at the Nordisk Konferanse Omarktisk Forskning, Ny Alesund, Svalbard. 2–8 August. 30 pp.

– 1986. "A Current Assessment of Cumulative Assessment." In *Cumulative Environmental Effects: A Binational Perspective.* Ottawa: Supply and Services Canada. Pp. 149–60

Rosenberg, D.M., et al. 1981. "Recent Trends in Environmental Impact Assessment." *Canadian Journal of Fisheries and Aquatic Sciences* 38:591–624

Rowley, G.W. 1966. "International Scientific Relations in the Arctic." In R.St.J. Macdonald (ed.). *The Arctic Frontier.* Toronto: University of Toronto Press. Pp. 279–92

Sadler, H.E. 1976. *The Flow of Water and Heat through Nares Strait.* DREO Report No. 736, Department of National Defence, Ottawa. 184 pp.

Sagoff, M. 1982. "We Have Met the Enemy and He Is Us, or Conflict and Contradiction in Environmental Law." *Environmental Law* 12:283–315

Salomonsen, F. 1979. "Trettende forelobige liste over genfundne gronlandske ringugle." *Dansk ornitologiske Forenings Tidsskrift* 73:191–206

Schwerdtfeger, W. 1963. "The Thermal Properties of Sea Ice." *Journal of Glaciology* 4:789–807

Science Council of Canada. 1975. *Canada's Energy Opportunities.* Report No. 23. Ottawa

– 1977a. *Northward Looking: A Strategy and a Science Policy for Northern Development* Report No. 26. Ottawa. 95 pp.

– 1977b. *Proceedings of the Seminar on Natural Gas from the Arctic by Marine Mode.* Ottawa

– 1979. *Roads to Energy Self Reliance – the Necessary National Demonstrations.* Ottawa

Scoresby, W. 1820. *An Account of the Arctic Regions, with a History and Description of the Northern Whale Fishery.* London: David and Charles reprint, 1969

Semtner, A.J., Jr. 1976. "A Model for Thermodynamic Growth of Sea Ice in Numerical Investigations of Climate." *Journal of Physical Oceanography* 6:379–89

Serson, H.V. 1972. *Investigation of a Plug of Multi-Year Ice in the Mouth of Nansen Sound.* Technical Note 72–6, Defence Research Establishment, Ottawa

Slovic, P., Fischhoff, B., and Lichtenstein, S. 1980. "Facts and Fears: Understanding Perceived Risk." In R.C. Schwing and W.A. Albers, Jr. (eds.). *Societal Risk Assessment.* New York: Plenum. Pp. 181–214

Smalley, R.D. 1980. "Risk Assessment: An Introduction and Critique." *Coastal Zone Management Journal* 7(2–3–4):133–62

Smiley, B.D. 1982. "The Effects of Oil on Marine Mammals." In J.B. Sprague, J.H. Vandermeulen, and P.G. Wells (eds.). *Oil and Dispersants in Canadian Seas – Research Appraisal and Recommendations.* Environmental Protection Service Report EPS 3–EC-82-2. Pp. 113–23

Smiley, B.D., and Milne, A.R. 1979. *LNG Transport in Parry Channel: Possible Environmental Hazards*. Sidney, BC: Institute of Ocean Sciences. 47 pp.

Smith, E.H. 1931. *The 'Marion' Expedition to Davis Strait and Baffin Bay under the Direction of the U.S. Coast Guard, 1928*. US Treasury Department Bulletin No. 19, Washington, DC

Smith, E.H., Soule, F.M., and Mosby, O. 1937. *The 'Marion' and 'General Green' Expeditions to Davis Strait and Labrador Sea, under Direction of the United States Coast Guard, 1928–1931–1933–1934–1935*. Pt. 2. *Scientific Results*. Washington, DC

Smith, M., and Rigby, B. 1981. *Distribution of Polynyas in the Canadian Arctic: Polynyas in the Canadian Arctic*. CWS Occasional Paper 45, 7–28

Smith, S.D., and Banke, E.G. 1981. "A Numerical Model of Iceberg Drift." POAC 2:1001–11

– 1983. "The Influence of Winds, Currents and Towing Forces on the Drift of Icebergs." *Cold Regions Science and Technology* 6:241–55

Smith, T.G., and Stirling, I. 1975. "The Breeding Habitat of Ringed Seals (PHOCA HISPIDA): The Birth Lair and Associated Structures." *Canadian Journal of Zoology* 53:1297–1305

Sowls, A.L., DeGrange, A.R., Nelson, J.W., and Lester, G.S. 1980. *Catalog of California Seabird Colonies*. US Department of the Interior, US Fish and Wildlife Service, Biological Services Program Report FWS/OBS–80/37, Washington, DC. 371 pp.

Sowls, A.L., Hatch, S.A., and Lensink, C.A. 1978. *Catalog of Alaskan Seabird Colonies*. US Department of the Interior, Fish and Wildlife Service, Biological Services Program Report FWS/OBS–78/78, Washington, DC

Stewart, R.W. 1971. *Ad Mare: Canada Looks to the Sea*. Ottawa: Science Council of Canada. 173 pp.

Stirling, I., and Calvert, W. 1983. "Environmental Threats to Marine Mammals in the Canadian Arctic." *Polar Record* 21(134):433–49

Stowe, T.J., and Underwood, L.A. 1984. "Oil Spillages Affecting Seabirds in the United Kingdom, 1966–1983." *Marine Pollution Bulletin* 15:147–52

Straughan, D. (ed.). 1971. *Biological and Oceanographical Survey of the Santa Barbara Channel Oil Spill 1967–1970*. Vols. I and II. Los Angeles: Allan Hancock Foundation, University of Southern California

Szaro, R.C., and Albers, P.H. 1977. "Effects of External Applications of No. 2 Fuel Oil on Common Eider Eggs." In D.A. Wolfe (ed.), *Fate and Effects of Petroleum Hydrocarbons in Marine Ecosystems and Organisms: Proceedings of a Symposium, Seattle, Washington, 1976*. New York: Pergamon Press. Pp. 164–7

Tanis, J.J.C., and Morzer Bruyns, M.F. 1968. "The Impact of Oil Pollution on Seabirds in Europe." *Proceedings of the International Conference on Oil Pollution of the Sea, 7–9 October, 1968, Rome*. Pp. 69–74

Task Force on Northern Conservation. 1984. *Report of the Task Force on Northern Conservation.* Ottawa: DIAND. 48 pp.

Task Force to Review Comprehensive Claims Policy. 1985. *Living Treaties, Lasting Agreements: Report of the Task Force to Review Comprehensive Claims Policy.* Ottawa: DIAND. 132 pp.

Tener, J. 1985. "Some Thoughts of Implementing Beaufort Sea Panel Recommendations." In P. Seely (ed.). *Making the Frontiers Conventional: Proceedings of the Fourteenth Annual APOA/CPA – ODD Workshop.* Calgary: Pallister Resource Management Ltd. Pp. B-22-29

Terhune, J. 1985. "Marine Survival." *Policy Options* (May): 23-5

Thomas, K., and Otway, H.J. 1980. "Public Perceptions of Energy System Risks: Some Policy Implications." In T. O'Riordan and K. Turner (eds.). *Progress in Resource Management and Environment Planning.* Vol. II. Toronto: John Wiley. Pp. 109-31

Thomson, A. 1948. "The Growth of Meteorological Knowledge of the Canadian Arctic." *Arctic* 1(4):34-43

Topham, D.R., Perkin, R.G., Smith, S.D., and Anderson, R.J. 1983. "An Investigation of a Polynya in the Canadian Archipelago." *Journal of Geophysical Research* 88(5):2888-916

Transche, N.A. 1928. "The Ice Cover of the Arctic Seas, with a Genetic Classification of Sea Ice." In *Problems of Polar Research.* American Geographical Society Publication No. 7 Pp. 91-123

Transport Canada. 1981. *Composite Executive Summary to the Independent "TERMPOL CODE" Assessments of the Three Arctic Pilot Project Sites Submitted.* 26 pp.

– 1985. *Arctic Marine Transportation R&D Plan, 1986/87 to 1990/91.* TP 7188E. Ottawa. 102 pp.

Tribe, L.H. 1974. "Ways Not to Think About Plastic Trees: New Foundation for Environmental Law." *Yale Law Journal* 83:1315.

Tuck, L.M. 1961. *The Murres.* CWS Monograph Series No. 1, Ottawa. 260 pp.

– 1971. "The Occurrence of Greenland and European Birds in Newfoundland." *Bird-Banding* 42:186-209

Tuck, L.M., and Lemieux, L. 1959. "The Avifauna of Bylot Island." *Dansk ornitologisk Forenings Tidsskrift* 53:137-54

Tuck, L.M., and Squires, H.J. 1955. "Food and Feeding Habits of Brunnich's Murre (*Uria lomvia lomvia*) on Akpatok Island." *Journal of the Fisheries Research Board of Canada* 12:781-92

Tungavik Federation of Nunavut. 1986. *Nunavut: A Report on Land Claims from the Tungavik Federation of Nunavut.* 8 pp.

United States, Interagency Arctic Research Policy Committee. 1987. *United States Arctic Research Plan.* NSF 87-55, Washington, DC

– 1988. *Arctic Research of the United States.* Vol. II. Washington, DC: National Science Foundation

Untersteiner, N. 1964a. "A Monograph for Determining Heat Storage in Sea Ice." *Journal of Glaciology* 5(39):352

– 1964b. "Calculations of Temperature Regime and Heat Budget of Sea Ice in the Central Arctic." *Journal of Geophysical Research* 69(22):4755–66

Van Ieperen, M.P. 1981. *Oceanographic Summary Report of Current, Tide, Temperature and Salinity Data (1974–1980).* Calgary: Panarctic Oils Ltd

Vandermeulen, J.H. 1978. "Self-Cleaning and Biological Recovery of an Oiled Marine Environment: ARROW 1970–1977." *Oceanus* 20:31–9

– 1980. "Introduction." In J. Vandermeulen (ed.). *Scientific Studies during the 'Kurdistan' Tanker Incident: Proceedings of a Workshop.* Bedford Institute of Oceanography Report Series No. BI-R-80-3, Dartmouth, NS. Pp. 5–7

– 1982. "Oil Spills: What Have We Learned?" In J.B. Sprague, J.H. Vandermeulen, and P.G. Wells (eds.). *Oil and Dispersants in Canadian Seas – Research Appraisal and Recommendations.* Environmental Protection Service Report EPS 3-EC-82-2. Pp. 29–46

VanderZwaag, D., Beanlands, G.E., Duinker, P.N., Taylor, R., and Underwood, P. 1986. "Decisionmaking Improvements and Alternatives." In A. Rieser, J. Spiller, and D. VanderZwaag (eds.). *Environmental Decisionmaking in a Transboundary Region: Fundy Tidal Power and the New England Coast.* New York: Springer-Verlag. Pp. 155–182

VanderZwaag, D., and Lamson, C. 1986. "Northern Decision Making: A Drifting Net in a Restless Sea." In C. Lamson and D. VanderZwaag (eds.). *Transit Management in the Northwest Passage: Problems and Prospects.* Cambridge: Cambridge University Press. Pp. 153–250

Veitch, C.R. 1978. "Seabirds Found Dead in New Zealand in 1976." *Notornis* 25:141–9

Vermeer, K. 1976. "Colonial Auks and Eiders as Potential Indicators of Oil Pollution." *Marine Pollution Bulletin* 7:165–7

Vermeer, K., and Peakall, D.B. 1977. "Toxic Chemicals in Canadian Fish-Eating Birds." *Marine Pollution Bulletin* 8:205–10

Wadhams, P. 1976. "Oil and Ice in the Beaufort Sea." *Polar Record* 18(114): 237–50

– 1977. "Characteristics of Deep Pressure Ridges in the Arctic Ocean." POAC 1:544–55

Waldichuk, M. 1980. "Retrospect of the Ixtoc I Blowout." *Marine Pollution Bulletin* 11:184–6

– 1982. "An Environmental Assessment and Review Process." *Marine Pollution Bulletin* 13:405–8

Walker, E.R. 1971. Monthly Mean Water Levels in the Canadian Arctic. Unpublished manuscript

Walker, E.R., and Lake, R.A. 1973. "Runoff in the Canadian Arctic Archipelago." In G.E. Weller (ed.). *Climate in the Arctic: Proceedings, Twenty-fourth Alaska Science Conference.* Pp. 374–8

Walker, E.R., and Wadhams, P. 1979. "Thick Sea Ice Floes." *Arctic* 32(2): 140–7

Walter, H., and Lien, J. 1985. *Attitudes of Canadian Students and Teachers toward the Marine Environment and Marine Education.* Prepared by Newfoundland Institute for Cold Ocean Sciences and Department of Psychology, Memorial University of Newfoundland, St John's, for DFO, Ottawa. 16 pp.

Wardley-Smith, J. 1983. "The Dispersant Problem." *Marine Pollution Bulletin* 14:245–9

Washburn, A.L. 1948. "International Cooperation in Arctic Research." *Arctic* 1(1):4–7

– 1980. "Focus on Polar Research." *Science* 209(4457):643–52

Watts, N., and Wandesforde-Smith, G. 1981. "Postmaterial Values and Environmental Policy Change." In D.E. Mann (ed.). *Environmental Policy Formation.* Lexington: Lexington Books. Pp. 29–42

Webster, N. 1977. *Webster's New Twentieth Century Dictionary.* 2nd ed. William Collins and World Publishing. 2129 pp.

Weeks, W.F., Kovacs, A., and Hibler, W.D. 1971. "Pressure Ridge Characteristics in the Arctic Coastal Environment." POAC 1:152–83

Weller, G.E. 1969. "Radiation Diffusion in Antarctic Ice Media." *Nature* 221:355–6

Wendt, S., and Cooch, F.G. 1984. *The Kill of Murres in Newfoundland in the 1977–78, 1978–79 and 1979–80 Hunting Seasons.* CWS Progress Note No. 146. 10 pp.

Wengert, N. 1985. "Citizen Participation: Practice in Search of a Theory." *Natural Resources Journal, 25th Anniversary Anthology* 68–85

Western Report. 1986. "The Icebreaker or Bust." 24 Nov, p. 12

White, D.H., King, K.A., and Coon, N.C. 1979. "Effects of No. 2 Fuel Oil on Hatchability of Marine and Estuarine Bird Eggs." *Bulletin of Environmental Contamination and Toxicology* 21:7–10

Whittington, M.S. 1985. "Political and Constitutional Development in the NWT and Yukon: The Issues and the Interests." In M.S. Whittington (ed.). *The North.* Toronto: University of Toronto Press. Pp. 53–108

Wilson, J.T. 1973. "International Research." In *Science and the North, a Seminar on Guidelines for Scientific Activities in Northern Canada, 15–18 October, Mont Gabriel, Quebec.* Ottawa. Pp. 236–59

Wolcott, D.M. 1980. *Presentation to Arctic Pilot Project EARP Panel Resolute Bay Technical Hearing, 23–29 April, Resolute Bay.* Calgary: Petro-Canada. 8 pp.

Wondolleck, J. 1985. "The Importance of Process in Resolving Environmental Disputes." *Environmental Impact Assessment Review* 5:341–56

Wynne-Edwards, V.C. 1935. "On the Habits and Distribution of Birds in the North Atlantic." *Proceedings of the Boston Society for Natural History* 40:233–346

Index

Sea Hydrocarbon Pro-
duction and Trans-
portation Proposal); Ca-
nadian Wildlife Service
(cws) role in, 94–5;
as conflict resolution
mechanism, 168; criteria
for significance and ac-
ceptability, 237; differ-
ent perspectives, 225;
EARP panels: composi-
tion and mandate of,
169, historical overview
of reports of, 86, les-
sons for future EARPS,
180–7, use of technical
specialists, 209; ensuring
judicial review, 238–9;
future of, in Canada,
97–100, 242, possible
improvements, 209–10,
220, 231–9, toward in-
dependent assessment,
233–4, toward policy
and program analysis,
232–3; Guidelines Order
(1989) legal status,
243–4; Lancaster Sound
exploratory drilling
EARP, 128; oil-spill un-
certainties, 98–100; pub-
lic participation in,
234–5, 238, public dis-
closure of information,
235–6; research needs,
long-term, 57, con-
straints of short-term
approaches, 95–7,
99–100
environmental impact as-
sessment (EIA), 20–1,
157–8 (see also Depart-
ment of Indian Affairs
and Northern Develop-
ment; risk assessment);
"adaptive environmen-
tal assessment and
behaviour," 16, 83,
Valued Ecosystem
Components (VECS), 83;
CCREM recommenda-
tions, 233; complexity

of, in Arctic, 157, 200,
201; criteria for deci-
sion-making, 179; data
assessment, 54–8, data
constraints and chal-
lenges, 55–7; matrix
technique, 201, post-
project monitoring,
186, 208, proprietary
information, 131, scop-
ing, critical view of,
184–5, types of assess-
ment, 54–5; discretion-
ary powers of review,
116; effectiveness and
efficiency of, 182–3;
GNWT's changing role
in, 154–7, case studies
of GNWT role in EIA,
158–65, GNWT perspec-
tives, 165–6; history of,
xi; impact management,
185–6; importance of
government manage-
ment of EIA, 182; indus-
try concerns, 103, 108,
109, 126–31; integrating
and improving scientific
input, 179–86, 228–31;
Inuit views, 147–53, on
Beaufort EARP, 142–7,
criteria for EIA, 151–3,
land claims and EIA, xii,
political aspirations
and EIA, 148, 150; mis-
use of, by Canadian
government, 150–1;
public partici-
pation (see participation,
public); recommenda-
tions for, 130–1, 166,
242–3
Environmental Studies
Revolving Fund (ESRF),
15
Esso Petroleum Ltd, 14,
61, 104
ethics, and environmental
management in the Arc-
tic, 220–6
exploration, arctic, pre-
1939 history of, 4–8

exploratory drilling: arctic
moratorium (1973), 14;
study of effects of, in
Canada, 14–16

Federal Environmental
Assessment Review Of-
fice (FEARO), 16, 60, 86,
93; and Canadian Envi-
ronmental Assessment
Research Council
(CEARC), 229; role of, in
EARP process, 169
Federal Court of Canada,
Trial Division, legal sta-
tus of EARP Guidelines
Order (1989), 243–4
fish, effects of spilled oil,
203
Fish and Wildlife Service,
GNWT, 159
Fisheries Act, 109–10; and
public participation,
223
Fisheries Research Board,
Canada, 9
Franklin expeditions, 5,
19, 105; role of, in pro-
moting research, 5
Frozen Sea Research
Group, Institute of
Ocean Sciences, 11, 14

Geological Survey of Can-
ada, 6
geostrophic water flow,
causing water-level var-
iations, 22
Godthaab expedition, 7
government of Northwest
Territories (GNWT), 108,
119, 122, 140; changing
role in EIA, 154–7; land
use planning process,
239; and various envi-
ronmental reviews,
158–65, APP review, 119,
122, 158–60, 161–3,
Beaufort Sea Proposal
review, 163–5; views on
EIA, 165–6
grain shipping, arctic,

Contributors

GORDON BEANLANDS is director, School for Resource and Environmental Studies, Dalhousie University, Halifax.

RICHARD G.B. BROWN is a research scientist with the Seabird Research Unit, Canadian Wildlife Service, Bedford Institute of Oceanography, Dartmouth, Nova Scotia.

JOHN DONIHEE is legal counsel for the Department of Justice with the Department of Renewable Resources, Government of the Northwest Territories, Yellowknife.

ROBERT L. DRYDEN is supervisor, Environment, for Chevron Canada Resources, Calgary.

DOUGLAS M. JOHNSTON is a professor of law and holds the Chair in Asia Pacific Legal Relations at the University of Victoria.

PETER JULL is a consultant, lecturer, and writer on circumpolar affairs, currently residing in Upper Coomera, Australia.

ROBERT A. LAKE is manager, Research Support, at the Institute of Ocean Sciences, Sidney, British Columbia.

CYNTHIA LAMSON is associate director, Oceans Institute of Canada / Institut canadien des océans, Halifax, and assistant professor (research) at the School for Resource and Environmental Studies, Dalhousie University, Halifax.

RAY LEMBERG is a consultant in risk analysis and risk management, based in Montreal.

DAVID W.I. MARSHALL directs the Pacific Region, Federal Environmental Assessment and Review Office, Vancouver.

HEATHER MYERS is regional planning analyst for the Department of Renewable Resources, Government of the Northwest Territories, living in Pond Inlet.

BRIAN SMILEY is marine adviser, Data Assessment Division, at the Institute of Ocean Sciences, Sidney, British Columbia.

DAVID L. VANDERZWAAG is director of the Marine and Environmental Law Program, Faculty of Law, Dalhousie University, Halifax. He is also a research associate at the Oceans Institute of Canada / Institut canadien des océans, Halifax.